Laboratory Instruments and Techniques Series

Consultant Editors:
M.M.Breuer MSc, PhD, FRIC
Principal Scientist, Unilever Research, England

A.D.Jenkins PhD, DSc, FRIC
Reader in Chemistry, University of Sussex

pH Meters

A.Wilson BSc, PhD, DIC
Unilever Research

Kogan Page London
Barnes and Noble New York

Copyright © 1970 **A.Wilson**

Designed by Laing Livesey Partners

Printed in Great Britain by
Lewis Reprints Limited
Port Talbot, Glamorgan

SBN 85038 130 4

pH Meters

PREFACE

In 1909 Sørenson defined the pH unit and outlined an experimental method for the determination of pH values. Since then both the theory and practice of pH measurement have changed. Standard procedures have been established and pH determinations are now among the physico-chemical measurements most frequently performed in chemical and non-chemical laboratories. In process and quality control pH measurement and regulation are often indispensible factors.

This book deals with the measurement of pH values and their meaning and use. Theoretical aspects have been considered only as a means of explaining the significance and limitations of experimentally obtained pH values. Basic experimental procedures have been described in some detail and the last chapter deals with various electrode systems and their applications.

Various types of pH meter are now commercially available, some of them for highly specific purposes. This book aims to enable the reader to select the most suitable pH meter for any particular problem. The characteristics of a large number of commercially available instruments are listed at the end of the book; these tables have been compiled from information supplied by the manufacturers.

A. Wilson
June 1970

List of symbols used

a_{H^+} = activity of hydrogen ion

E_x = EMF of cell using solution X

E_s = EMF of cell using standard solution S

E_0 = standard potential at equilibrium

E_b = EMF of cell associated with cell reaction

E_l = EMF of cell associated with liquid junction

f_i = activity coefficient of component, in mol. fraction^{-1} units

ΔG = change in free energy

I = ionic strength

K_W = ionic product of water

K'_W = concentration ionic product

K^a_{HA} = dissociation constant of acid HA

K^b_{Ac-} = basic dissociation constant of Ac^-

l_{st} = length on potentiometer required by standard cell

l_m = length on potentiometer required by measuring cell

n_x = no. of molecules of component X

pH = $-\log_{10} a_H$

pcH = $-\log_{10} H^+$

pK^a_{HA} = $-\log K^a_{HA}$

$\Delta pH_{\frac{1}{2}}$ = dilution value of buffer

R_{pH} = pH response of glass electrode

S = entropy

V = volume

Z = ionic charge

α = degree of ionisation

β = buffer capacity

β_e = electromotive efficiency of glass electrode

μ_x = partial molal free energy or chemical potential of component X

γ = activity coefficient in molality^{-1} units

Tables

Appendices

CONTENTS

The calomel reference electrode
The silver–silver chloride electrode
The thallium amalgam-thallium (I) chloride electrode
Ion-specific electrodes

1. pH AND pH MEASUREMENT

1.1. What is pH ?

The acidity or basicity of an aqueous solution depends on the concentration of free hydrogen (H^+) ions and hydroxyl (OH^-) ions present in it. In pure water, which is exactly neutral, the concentration of free hydrogen ions is equivalent to the concentration hydroxyl ions. In solutions of acids (defined by Brønsted and Lowry as proton donors), hydrogen ions predominate while in basic solutions the reverse is true.

The dissociation of water is written

$$H_2O \rightleftharpoons H^+ + OH^-$$

The *concentration ionic product of water* K'_w is defined as

$$K'_w = [H^+][OH^-]$$

In pure water $[H^+] \approx 10^{-7}$ grammes per litre

and also $[H^+] \equiv [OH^-]$

$$\therefore K'_w = [H^+][OH^-] \approx 10^{-14}$$

The pH of a solution is a measure of its 'acidity' or basicity. It must therefore be a function of the concentration of hydrogen ion present in it. Sorenson, who put forward the concept of pH, defined it as follows:

$$pH = -\log_{10}[H^+] = \log_{10}\frac{1}{[H^+]}$$

or $[H^+] = 10^{-pH}$

pH values were thus the logarithms to the base ten of the reciprocal of hydrogen ion concentrations in grammes per litre.

In pure water where

$[H^+] \approx 10^{-7}$ grammes per litre

$pH \approx 7$

In a decinormal hydrochloric acid solution where dissociation can be taken to be complete

$[H^+] = 10^{-1}$

and $pH = 1$

Similarly, in decinormal sodium hydroxide solution

$[OH^-] = 10^{-1}$ gramme ions per litre

Since $[H^+] = \dfrac{K'_w}{[OH^-]}$

$\qquad [H^+] = 10^{-13}$

and $\therefore$ pH = 13

With the development of chemical thermodynamics it became evident that Sørenson's experimental pH values did not yield the exact hydrogen ion concentration. They were in fact functions of the apparent concentrations (as determined by other physical properties) of the electrolytes in solution. This apparent concentration is known as *activity*, the ratio activity concentration being termed the *activity coefficient*. Hydrogen ion activity has now replaced ion concentration in the definition of pH:

$$pH = -\log_{10}(a_{H^+})$$

where a_{H^+} is the activity of the hydrogen ion.

1.2. Measurement of pH

There are two main methods of measuring the pH of a solution – the *colorimetric* or *indicator method* and the *electrometric* or *pH meter* method.

An indicator is a dye whose colour is determined by the acid-base ratio of the solution to which it is added. It behaves like a weak acid or base, the acid form giving a colour different from the basic form. At any pH, both forms are present, but the ratio of their concentrations and hence the colour produced is dependent on the hydrogen ion concentration. Colorimetric determinations give only approximate results but they have the advantage of speed and low cost.

Electrometric methods are far more accurate. They involve the determination of the EMF generated by a galvanic cell which can be represented by: R | soln. R ⦙ $\mathcal{J}$ ⦙ soln. X | M where R is a reference electrode placed in a solution R, M is a measuring electrode immersed in X (the solution where pH is being determined) and $\mathcal{J}$ (which may be a liquid junction) is a means of contact between the two electrodes.

The pH of the solution X is a linear function of the measured EMF, Ex

$$pH(X) = \text{constant} + \frac{E_x}{2\cdot3026\text{RT}/\text{F}} \qquad \ldots i$$

For a standard solution (S)

$$\text{pH}(S) = \text{constant} + \frac{E_s}{2\cdot3026\text{RT}/\text{F}} \qquad \ldots ii$$

Subtracting ii from i gives

$$\text{pH}(x) = \text{pH}(S) + \frac{E_x - E_s}{2\cdot3026\text{RT}/\text{F}} \qquad \ldots iii$$

This is the equation commonly used in a pH determination. The standard solution is one which has been selected and assigned a pH by British Standard 1647:1961 or the National Bureau of Standards (see "the pH Scale" Chapter II).

The electric circuit required for pH measurement must allow the potential difference between the two electrodes to be measured when there is no flow of current through the cell. One such arrangement, the Poggendorff compensation method, is shown in figure 1.

'St' is a Weston Standard cell and M is the pH-measuring cell. XY is a length of uniform resistance wire on which the position of the sliding contact S is adjusted so that there is no deflection of the galvanometer G, first when a and c are connected (reading l_{st}) and then when b and c are connected (reading l_M). The EMF of the Weston Standard cell and the measuring cell are then related as follows:

$$\frac{E_{st}}{E_M} = \frac{l_{st}}{l_M}$$

E_M can be obtained since the other three quantities are known. E_s in equation iii can be similarly measured and since pH (S) is known, pH (x) can be evaluated.

The Poggendorff method is insensitive for cells with a high internal resistance (e.g. those using a glass electrode). Most pH meters now have electronic circuits usually involving electrometer valves. A simplified version of an electronic potentiometer circuit used in some commercial pH meters is shown in figure 2. X and Y are first connected and the rheostat P adjusted to give a suitable zero reading of the galvanometer G. X and Z are then connected and the sliding wire adjusted till a null reading is again obtained and the value of E_M can then be calculated.

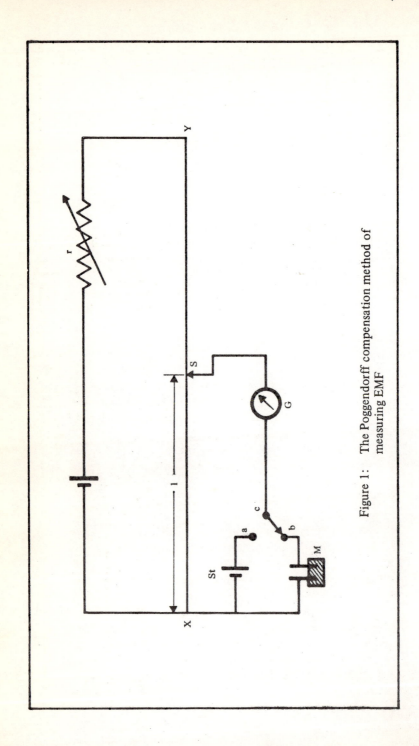

Figure 1: The Poggendorff compensation method of measuring EMF

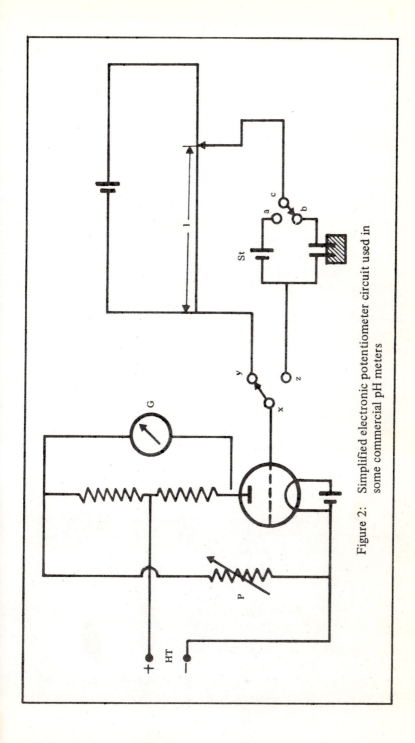

Figure 2: Simplified electronic potentiometer circuit used in some commercial pH meters

2. BASIC PRINCIPLES OF pH MEASUREMENT

The electrometric determination of pH consists essentially of the evaluation of a thermodynamic constant from a measurement of the electromotive force of a suitable electrochemical cell. In this section we shall briefly consider the cell process and the thermo-dynamics of electrolytic solutions, these basic principles being essential for an understanding of the theory and practice of pH measurement.

2.1. Electrochemical cells and their free energy

When current is drawn from a galvanic cell a chemical process called *cell reaction* takes place and chemical energy is transformed into electrical energy. Electrons are produced, and oxidation occurs at one electrode. At the other electrode electrons are utilised and reduction takes place. A reversible electrode can act as a site for either of these two reactions. The *potential*, or tendency for a given electrode to supply electrons, depends on the free energies of the oxidized and reduced states of the chemical substances present, the temperature and the pressure. The electromotive force (EMF, E) is the driving force of the reaction.

The EMF of a galvanic cell is measured by the compensation method, i.e. the EMF of the cell is balanced almost exactly by that supplied by the potentiometer. In this situation the condition of thermodynamic reversibility is closely approached and the principles of equilibrium thermodynamics become applicable.

Let us consider the cell

$$H_2(Pt) \mid H^+, Cl^-(aq) \mid Hg_2Cl_2 \mid Hg$$

Here the 'oxidizing agent' mercury (I) chloride is isolated from the reducing agent, the hydrogen atoms. For cell reaction to occur a transfer of electric charge must take place; when there is no such transfer, the reaction at the electrodes ceases. However, the tendency for the cell reaction:

$$\tfrac{1}{2}H_2(gas) + \tfrac{1}{2}Hg_2Cl_2(insoluble) + H_2O(solvent)$$
$$\rightarrow H_3O^+(solute) + Cl(solute) + Hg(insoluble) \ldots 1$$

to occur still exists and is revealed by the difference in EMF between the platinum and mercury electrodes.

The decrease in free energy ΔG which accompanies the reaction is equivalent to the maximum electrical work produced per unit chemical reaction at constant temperature and pressure. This can be evaluated as the reversible EMF, E, multiplied by the quantity of electric charges required to be transferred for unit chemical reaction to take place

$$-\Delta G = nFE \qquad \qquad \ldots 2$$

(F is the Faraday and n an integer, if E is expressed in volts, ΔG is obtained in Joules.)

Chemical potential and activity

The free energy change ΔG like the EMF, is a measure of the tendency for the cell reaction to take place.

Let us now consider the cell reaction

$$iI + j\mathcal{J} + \ldots = uU + vV + \ldots \qquad \qquad \ldots 3$$

representing the conversion of i moles of component I, j moles of $\mathcal{J}$, etc., into u moles of U, v moles of V, etc. When there is a small change in one or more of the variables, temperature, pressure and composition, the change in the free energy can be expressed as follows:

$$dG = -SdT + VdP + \Sigma \mu_x dn_x \qquad \qquad \ldots 4$$

where
n_x = the number of moles of component x,
S and V are the entropy and volume of the system respectively and μ_x is a quantity called the *partial molal free energy* or *chemical potential* of species x, defined as follows:

$$\mu_x = \left(\frac{\delta G}{\delta n_x}\right)_{T,P,n} \qquad \qquad \ldots 5$$

μ_x represents the changes of free energy at constant temperature and pressure caused by the reversible addition of one mole of species x to such a large amount of the system that the change in concentration is negligible.

It is often convenient to express the chemical potential of each species in any particular state in terms of the activity 'a' of that species thus:

$$\mu_x - \mu_x^0 = RT \, ln a_x \qquad \qquad \ldots 6$$

Here R is the gas constant and μ_x^0 is the chemical potential of x in the standard or reference state. The reference state for a liquid solution is an infinitely dilute solution. For gases the reference state gives a value of activity equal to the partial pressure of the component. Pure solids and liquids at atmospheric pressure are in their reference states.

When reaction 3. occurs at constant temperature and pressure the reaction interval is small enough to assume that the chemical potentials of the reactants and the products remain constant, we get from eqns. 4. and 6.

$$\Delta G = u\mu_U + v\mu_V + \ldots - i\mu_I - j\mu_J \qquad \ldots 7$$

$$\Delta G = \Delta G^0 + RT \ln \frac{a_U^u \, a_V^v}{a_I^i \, a_J^j} \qquad \ldots 8$$

where ΔG^0 = change in free energy when all substances are in their standard states.

i.e. $$\Delta G^0 = u\mu_U^0 + v\mu_V^0 + \ldots - i\mu_I^0 - j\mu_J^0 \qquad \ldots 9$$

From equations 2. and 8. we get two important expressions: first, that

$$E = E_0 - \frac{RT}{nF} \ln \frac{a_U^u \, a_V^v}{a_I^i \, a_J^j} \qquad \ldots 10$$

and second, that

$$\Delta G^0 = -RT \ln \frac{a_U^u \, a_V^v}{a_I^i \, a_J^j} \ldots = -RT \ln K| \qquad \ldots 11$$

K is the equilibrium constant of the cell reaction and the activities are values at equilibrium. This is the form which equation 8. takes in the equilibrium state at constant temperature and pressure, ΔG being equal to zero in such a situation.

It can be seen from these equations that E_0, the standard potential, relates to the equilibrium state

$$E_0 = \frac{RT}{nF} \ln K \qquad \ldots 12$$

and the EMF, E, becomes zero as equilibrium is attained.

So far the unit of activity has not been defined. We shall not discuss it at length here except to say that the choice of the unit is largely arbitrary, the numerical value of the activity being dependent on the standard state. This is demonstrated by equation 6. In the standard state in liquid solutions m_i, the molality, tends to zero and $a/m_i = 1$.

In real solutions a/m_i gives the value of the activity coefficient of the species

$$a/m_i = \gamma_i$$

The standard state may also be defined using concentration units of molarity C, or mole fraction (N), a/C_i or a/N_i becoming unity when Ci or Ni is zero. The activity coefficients in these two forms are termed y_i and f_i respectively.

Activity coefficients evaluated in electrolytic solutions are always mean values relating to combinations of ions. If the electrolyte of molality 'm' and activity 'a' dissociates into Z ions of which Z_+ are positive and Z_- negative, the mean ionic activity $a_{\pm}$ is given by:

$$a_{\pm} = (a_+^{z+}\, a_-^{z-})^{1/z} = a^{1/z} \qquad \ldots 13$$

The mean ionic molality is

$$m_{\pm} = m(z_+^{z+}\, z_-^{z-})^{1/z} \qquad \ldots 14$$

and the mean activity coefficient is

$$\gamma_{\pm} = \frac{a_{\pm}}{m_{\pm}} = (\gamma_+^{z+}\, \gamma_-^{z-})^{1/z} \qquad \ldots 15a \qquad \text{or}$$

$$f_{\pm} = (f_+^{z+}\, f_-^{z-})^{1/z} \qquad \ldots 15b$$

At low concentration the activity coefficients in aqueous solution vary mainly with interionic distances and the ionic charge; these factors are combined in the ionic strength I defined by Lewis and Randall as

$$I = \tfrac{1}{2}\Sigma m_i z_i{}^2$$

the summation being for all the ionic species present in solution.

In a solution containing 0·05 M KCl and 0·01 M K_2SO_4, for example, z^2 values for K^+, Cl^- and SO_4^{--} are $(1)^2$, $(-1)^2$ and $(-2)^2$ respectively and m_i has values of 0·07, 0·05, 0·01.
$$\therefore I = \tfrac{1}{2}[(0{\cdot}07 \times 1) + (0{\cdot}05 \times 1) + (0{\cdot}01 \times 4)] = 0{\cdot}08 \text{ M}.$$

2.2. The use of EMF for pH measurements

Let us consider again the cell

$$H_2(Pt) \| H^+, Cl^-(aq) \mid Hg_2\,Cl_2 \mid Hg$$

Using the cell reaction as represented by equation 1. we can write equation 10. in the form:

$$E = E_0 - \frac{RT}{F} \ln\left(\frac{[H^+][Cl^-]\, f_{H+}\, f_{Cl-}}{[H_2]^{\frac{1}{2}} \cdot f_{H_2O}} \right) \qquad \ldots 16$$

(solvent concentration and activities of insoluble phases being omitted) f_{H_2O}· expresses changes in solvent activity from the value for pure solvent, and this also can be omitted for dilute solutions, i.e., for ionic strength $<0\cdot1$. Under these conditions if we express hydrogen gas activity by its pressure equation 16. becomes:

$$E = E_0 + \frac{RT}{2F} \ln p - \frac{RT}{F} \ln[H^+][Cl^-] - \frac{2RT}{F} \ln f_{\pm}\, (H^+Cl^-)$$

$$\ldots 17$$

When HCl is the only solute present the concentrations of hydrogen and chlorine ions are both equal to [HCl], the stoichiometric concentration of HCl.

Equation 17. can then be used to evaluate E_0 by measuring E over a range of concentrations of HCl and calculating the sum

$$E^* = E - \frac{RT}{2F} \ln p + \frac{2RT}{F} \ln [HCl] \text{ (stoicheiometric)} \quad \ldots 18$$

for each measurement. As the solutions get more dilute

$$f_{\pm (H+Cl^-)} \to 1$$

and

$$\frac{2RT}{F} \ln f_{\pm (H+Cl^-)} \to 0$$

and hence $E^* \to E_0$

Unfortunately this straightforward treatment cannot be applied except to the simplest cells. Consider for example the very common type of cell represented by

$$M_1 \mid S_1 \,\vdots\, S_3 \,\vdots\, S_2 \mid M_2 \qquad \ldots 19$$

where M_1 and M_2 are metal electrodes immersed in solutions S_1 and S_2, connected by a third solution S_3. Here it is impossible to express in exact terms the chemical reaction occurring at the liquid interfaces (the broken vertical lines in 19.) involving migration and diffusion of species and causing changes in concentrations and activity coefficients. The complete cell reaction associated with the passage of one Faraday is consequently unknown.

Considering the problem in terms of EMF the total EMF E of a cell can be split into two main components

$$E = E_B + E_l \qquad \ldots 20$$

E_B is the EMF associated with chemical reaction at the electrodes while E_l the liquid junction potential is caused by the complex reactions at the liquid junctions.

In the cell:

$$\text{H}_2(\text{Pt}) \left| \begin{array}{c} H^+ \, X^- \text{(aqueous)} \\ \text{soln. } X \end{array} \right| \left. \begin{array}{c} \text{KCl (aqueous)} \\ \text{soln. } R \end{array} \right| \left. \begin{array}{c} \text{Hg}_2\,\text{Cl}_2 \end{array} \right| \text{Hg} \quad \ldots 21$$

which is of particular interest for pH measurement, the chemical reaction taking place at the electrodes on the passage of one Faraday through the cell from left to right is:

$\tfrac{1}{2}\text{H}_2$ gas (pressure $= p$) $+ \tfrac{1}{2}\text{Hg}_2\,\text{Cl}_2$ (insoluble)
$+ \, \text{H}_2\text{O}$ (solvent) $\rightarrow \text{H}_3\text{O}^+$ (in solution X)
$+ \, \text{Cl}^-$ (in solution R) $+ \, \text{Hg}$ (insoluble) $\qquad \ldots 22$

and the associated EMF is

$$E_B = (E_B)_0 - \frac{RT}{F} \, ln \, \frac{\{f_{\text{H}^+}[\text{H}^+]\}_X \, \{f_{\text{Cl}^-}[\text{Cl}^-]\}_R}{f_{\text{H}_2\text{O}} \cdot p^{1/2}} \qquad \ldots 23$$

Of these terms, the one obtained from experimental measurements is E and neither E_B nor E_l can be determined separately. Various methods have been tried to obtain E_B,

1. devising experimental arrangements for the liquid junctions which make E_l insignificantly small,
2. calculating E_l with the help of reasonable assumptions about the chemical reactions at the liquid junctions and,
3. combining measurements of E_B^0 in pairs and subtracting to eliminate E_l, provided E_l remains the same for both measurements. Each of these procedures has been developed to the point where results are accurate enough for significant information to be calculated from them, but none of them leads to an exact solution of the problem. The existence of the liquid junction in the cell sets a fundamental limitation to the significance of E_B values and of the derived information about pH in solutions.

Method 3. is the standard technique adopted in practical pH measurements using galvanic cells. Two cells of type 21. differing only in the nature of the solution surrounding the hydrogen electrode are used:

$$\text{H}_2(\text{Pt}) \quad \begin{array}{|c} \text{Solution } X \end{array} \begin{array}{c} \text{KCl (aqueous)} \\ \text{(3.5 N or saturated)} \\ \text{(Solution } R) \end{array} \begin{array}{c} \text{Hg}_2\text{Cl}_2 \end{array} \begin{array}{|c} \text{Hg} \end{array}$$

$$\text{H}_2(\text{Pt}) \quad \text{Pressure } p \qquad \qquad \qquad \qquad \qquad \dots 24$$

and

$$\text{H}_2(\text{Pt}) \quad \begin{array}{|c} \text{Solution } S \end{array} \begin{array}{c} \text{Solution } R \end{array} \qquad \begin{array}{c} \text{Hg}_2\text{Cl}_2 \end{array} \begin{array}{|c} \text{Hg} \end{array}$$

$$\text{Pressure } p \qquad \qquad \qquad \qquad \qquad \qquad \dots 25$$

Solution X is the unknown solution and solution S is the solution of standard pH with which it is being compared.

In accordance with equation 23. the corresponding EMFs are given by

$$(E_B)_X = (E_B)_0 - \frac{RT}{F} \, ln \, \frac{\{f_{\text{H}^+}[\text{H}^+]\}_X \, \{f_{\text{Cl}^-}[\text{Cl}^-]\}_R}{f_{\text{H}_2\text{O}}^X \cdot (p_X^{1/2})} + (E_l)_X$$

$$\dots 26$$

$$(E_B)_S = (E_B)_0 - \frac{RT}{F} \, ln \, \frac{\{f_{\text{H}^+}[\text{H}^+]\}_S \, \{f_{\text{Cl}^-}[\text{Cl}^-]\}_R}{f_{\text{H}_2\text{O}}^S \, (p_S^{1/2})} + (E_l)_S$$

Putting activity coefficients of water equal to unity, which is a reasonable assumption at constant temperature

$$(E_B)_X - (E_B)_S = \frac{RT}{F} \, ln \, \frac{\{f_{\text{H}^+}[\text{H}^+]\}_S}{\{f_{\text{H}^+}[\text{H}^+]\}_X} + \frac{RT}{2F} \, ln \frac{p_X}{p_S} + (E_l)_X - (E_l)_S$$

$$\dots 28$$

Since in practical measurement p_X is nearly equal to p_S the second term on the right hand side is only a small and easily evaluated correction and in later disuccion we shall omit it.

$E_l)_X - (E_l)_S$ is usually fairly small (within ± 2 mv) when ionic strength is <0.1. This is therefore the limit of errors in evaluation of the ratio of activities of hydrogen ions in solution S and X,

There is no term representing the reference electrode in equation 28. and experiments using different reference electrodes have given completely concordant results so that we can generalize cells 24. and 25. as follows:·

$$\text{H}_2(\text{Pt}) \quad \begin{array}{|c} \text{solution } X \\ \text{or } S \end{array} \begin{array}{c} \text{KCl aqueous} \\ \text{(3.5 N or} \\ \text{saturated} \\ \text{solution)} \end{array} \begin{array}{c} \text{reference} \\ \text{electrode} \end{array} \qquad \dots 29$$

2.3. The pH scale

In the previous section a method comparing the hydrogen ion activity of an unknown solution with that of a standard has been outlined. To obtain the pH of the unknown solution a reference scale (which assigns a pH value to the standard and finally to the unknown solution) is required. Unfortunately it was not until 1950 that any such scale was established. The standards now officially recognised in Britain and the United States are based on the work of standardization and recommendations of the U.S. National Bureau of Standards.

Both the British and U.S. authorities define a pH unit as the difference in the hydrogen ion activity of two standards which at 25°C, and using the same standard H_2 electrode (i.e., 1 atmos. H_2 pressure) and the same reference electrode, give an EMF difference of 0·05916 volts. The reference scale is then anchored by defining the nature of the standard solution and assigning a pH value to it. In the British Standard this is an 0·5 M solution of potassium hydrogen phthalate of high purity (free from sodium salt) with excess phthalic acid.

The pH of this solution at t°C is

$$pH(S) = 4 \cdot 000 + \tfrac{1}{2}\left(\frac{t-15}{100}\right)^2 \qquad \ldots 30$$

Appendix 2 gives some values of pH of this standard solution at different temperatures. The British standard pH scale applies to aqueous solutions between 0° and 60°C and is intended to be accurate within ±0·005.

The NBS recommendations define not one but a small number of standard solutions (of which the phthalate solution is one), the pH of each of these solutions having been determined with extreme care over a range of temperatures (Appendix 3). The British and NBS scales are entirely concordant. Measurements relating to the NBS scale are made on the cell

$$H_2(Pt) \mid \text{Solution } S + KCl \mid AgCl \mid Ag \qquad \ldots 31$$

which has no liquid junction. The reaction for the passage of one Faraday from left to right through the cell is

$$\tfrac{1}{2}H_2 \text{ (gas)} + AgCl \text{ (insoluble)} + H_2O \text{ (solvent)}$$
$$\rightarrow H_3O^+\text{(solute)} + Cl^-\text{(solute)} + Ag \text{ (insoluble)} \qquad \ldots 32$$

The only other standard of importance apart from the British and NBS ones is Japanese. This incorporates elements of both the

British and NBS methods and is entirely in accord with them. The scale applies between 0° and 60°C and the primary standard is 0·5 M phthalate solution. For initial adjustment of the pH meter a solution of 0·025 M (equimolal) phosphate buffer is used, the choice of the second standard buffer being dependent on the pH of the test solution. For a pH less than 2, 0·05 M solution of potassium tetra-oxalate is used; between pH 2 and 7 it is 0·025 phosphate, pH 7 to 10 uses a 0·01 M borax solution and pH 10 to 11 a solution 0·025 M with respect to both sodium bicarbonate and sodium carbonate.

2.4. pH and chemical equilibria

Most chemical processes involve either transfer of a proton or changes in the oxidation state and the activities of the reactants. These processes can be monitored by changes in the electro-chemical potential E, as shown in equation 10. (the Nernst Equation).

$$E = E_0 - \frac{RT}{nF} \, ln \, \frac{a_U^u \, a_V^a}{a_I^i \, a_J^j} \qquad \ldots 32$$

Let us consider an acid HA in aqueous solution taking part in the equilibrium

$$HA + H_2O \leftrightharpoons H_3O^+ + A^-$$

The acid dissociation constant is

$$K_{HA}^a = \frac{[A^-][H^+]}{[HA]} \, \frac{f_{A-} \, f_{H+}}{f_{HA}} \qquad \ldots 33$$

Neglecting activity coefficients equation 33. can be written in the following form

$$\log_{10} K_{HA}^a = \log_{10} \frac{[A^-]}{[HA]} + \log_{10} [H^+] \qquad \ldots 34$$

introducing the notation

$$pK_{HA}^a = -\log_{10} K_{HA}^a$$

and rearranging 34.

$$pK_{HA}^a = -\log_{10}[H^+] - \log \frac{[A^-]}{[HA]} \qquad \ldots 35$$

or

$$pH = pK_{HA}^a + \log\frac{[A^-]}{[HA]} \qquad \ldots 36$$

This is called the *Henderson–Hasselbalch equation*.

It can also be written in terms of α, the degree of dissociation (i.e., the fraction of molecules present in an ionized form)

$$pH = pK_{HA}^a + \log\frac{\alpha}{1-\alpha} \qquad \ldots 37$$

2.5. pH changes during acid-base titrations

Some general definitions

To avoid later ambiguities we shall first define some important terms commonly used in titration experiments.

Absolute neutrality
When the concentrations of hydrogen and hydroxyl ions are exactly balanced, the reaction is of *absolute neutrality*. In super-pure water, which is taken as the standard of absolute neutrality, the concentration of each ion at 18°C is $10^{-7.07}$ grammes per litre so that pH of the water is 7·07 (usually taken as 7·00).

Neutrality and neutral point
Although absolute neutrality is defined as above, the term *neutral* is frequently used in a different sense. A solution may be described as neutral to methyl orange or some other indicator. This means that the solution will give an intermediate colour with the particular indicator employed.

Equivalence point
This term is often used in order to avoid confusion caused by the terms neutral and neutral point. It describes a stage in the titration reaction where the reacting weights of acid and base are equivalent.

Acid-base titrations
As an acid is titrated against an alkali the pH of the solution undergoes a large change which is particularly rapid near the equivalence point. In order to follow this change, the titration vessel is made to constitute part of a cell suitable for pH measurements. This is done by immersing the measuring electrode in the solution to be titrated and joining the solution through a salt bridge to the reference half of the cell. The titrant and titrated solution

are stirred and EMF recorded after each small addition over the whole range.

Consider first the case of a strong acid being titrated against a strong base in aqueous solution. If 'C' is the normality of the acid (i.e., the concentration of hydrogen ion since ionization is complete for a strong acid) and if we add 'x' equivalents of a hypothetical infinitely concentrated solution of a strong base (i.e., x equivalents of hydroxyl ions), before the equivalence point:

$$[H^+] = C - x \qquad \qquad \ldots 38$$

At the equivalence point by definition

$$[H^+] = [OH^-] = (K'_w)^{\frac{1}{2}} \approx 10^{-7} \qquad \ldots 39$$

(K'_w being related to the ionic product of water by the relationship:

$$K_w = [H_3O^+][OH^-]f_{H_3O^+} f_{OH^-} = K'_w \, f_{H_3O^+} f_{OH^-})$$

After the equivalence point

$$[H^+] = \frac{K'_w}{[OH^-]} = \frac{K'_w}{x - C} \qquad \ldots 40$$

We can express the change in hydrogen ion concentration as a change in pcH where pcH $= -\log[H^+]$ (i.e., the negative logarithm of the H^+ ion concentration and not the H^+ ion activity which is pH). Figures 3 and 4 illustrate these changes for two types of titrations. Plots are of x/C against pcH for a strong and weak acid against a strong base.

The limits of accuracy accepted in an acidimetric titration are usually ± 0.1 per cent. To obtain this it is necessary that the pH change near the end point (i.e., from 99.9 per cent neutralized to 0.1 excess titrant) should be of the order of 1.6 pH units.

If the steep inflexion of the EMF curve is used as an indication of the equivalence point it is unnecessary to distinguish between pcH and pH, since in this region both quantities follow almost identical curves. The fact that the experimental curves are slightly different from those in figures 3 or 4 can also be neglected. (The difference is caused by dilution which inevitably accompanies addition of the titrant.)

Now consider the case of a weak acid being titrated against a strong base. Ionization of the acid in aqueous solution is incomplete and before the equivalence point there exists in solution

27

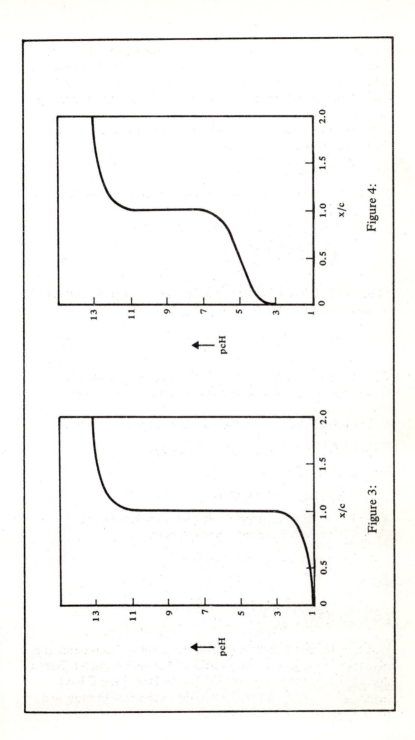

Figure 3:

Figure 4:

a mixture of weak acid and its salt. If we take acetic acid as the weak acid and sodium hydroxide as the strong base, at equivalence point we shall have a solution of pure sodium acetate only and beyond the equivalence point a mixture of the salt and increasing amounts of sodium hydroxide.

At the equivalence point, representing the acetate ion as Ac^- the extent of the reaction

$$Ac^- + H_2O \rightarrow HAc + OH^-$$

(i.e., the hydrolysis of salt NaA) will determine the value of hydrogen ion concentration.

The hydrolysis constant of NaAc is given by

$$K_{A^-}^b = \frac{[OH^-][HAc]}{[Ac^-]} \qquad \ldots 41$$

At the equivalence point HAc and OH^- are present in equivalent amounts and therefore

$$K_{A^-}^b = \frac{[OH^-]^2}{[Ac^-]} \qquad \ldots 42$$

When hydrolysis is negligible $[Ac^-]$ is practically equal to 'C', the stoichiometric concentration of the salt at the equivalence point.

Now using the definition for K_w, the ionic product of water:

$$K_w = [H_3O^+][OH^-] f_{H_3O^+} f_{OH^-} = K'_w \cdot f_{H_3O^+} f_{OH^-}$$

and the relationship

$$K_A^b \cdot K_{HA}^a = [H_3^+O][OH^-] f_{H_3O^+} f_{OH^-} = K_w \qquad \ldots 43$$

which is true for all conjugate acid-base pairs irrespective of charge, it is possible to derive the expression

$$pcH = \tfrac{1}{2} (pK'_w + pK'^a_{HA} + \log C) \qquad \ldots 44$$

where

$$pK'^a_{HA} = pK^a_{HA} \Big/ \frac{f_{H^+} f_{A^-}}{f_{HA}}$$

Equation 44. shows that the equivalence point will now occur at a pH greater than $\tfrac{1}{2}pK'_w$ provided the acid is weak enough to have a dissociation constant less than 10^{-3} moles litre^{-1} and C has a magnitude usual in analytical work. For such a case titrating at a pH 7 would not give a correct result.

2.6. Buffers

Buffers are solutions used to preserve a constant pH in experimental systems. There are two main types of buffer.

(a) Solutions containing a weak acid and its salt with a strong base, a weak base and its salt with a strong acid or a mixture of weak acid and weak base. The pH stability of such solutions is due to the presence of the weak acid or base ion.

(b) Solutions of strong acid or base of concentration greater than 0·01 M. pH stability in this case is caused by the logarithmic relationship between pH and concentration.

For approximate calculations the Henderson–Hasselbalch equation may be used to calculate the pH of buffer solutions of type (a). Thus for weak acids (HX) and their salts with strong bases (MX)

$$pH = pK_{HA}^a + \log \frac{C_{MX}}{C_{HX}}$$

The buffer capacity of an acid-base system is defined by

$$\beta = \frac{dc_{OH^-}}{dpH} \quad \text{or} - \frac{dc_{H^+}}{dpH}$$

When β is unity for a given solution, one gramme equivalent of hydrogen ions when added to one litre of the solution will cause a decrease of one pH unit; one gramme equivalent of hydroxyl ions will similarly cause an increase of one pH unit. When concentrations of acid and salt are equal β is at a maximum and in dilute solutions it is approximately equal to $2.303C/4$, C being the stoicheiometric concentration of the acid. The value of buffer capacity can be obtained from potentiometric titration curves like figure 4. At any point, β is the reciprocal of the slope of the curve.

Buffers need to be insensitive to dilution as much as to acid or base addition. The dilution value $\Delta pH_{\frac{1}{2}}$ is therefore an important function. It is the increase in pH of a buffer solution when it is diluted with an equal volume of water.

$$\Delta pH_{\frac{1}{2}} = (pH)_{\frac{1}{2}c} - (pH)_c \qquad \qquad \ldots 45$$

This effect is due to change in activity coefficients. It can cause difficulties in EMF measurements on dilute solutions.

Since buffer solutions are used mainly in the standardization of pH cell assemblies, the preparation of buffers is of the utmost importance. Manufacturers of pH instruments sometimes supply 'buffer tablets' which when dissolved in the specified amount of distilled or deionized water give buffer solutions according to BS or NBS specifications. The tablets may contain polyethylene glycol, dextrose, glucose or other substances which aid tableting and do not significantly alter the pH of the final solution. For accurate work, it is essential that buffer solutions be prepared with care according to standard methods.

Table 1
Properties of buffer systems*

Buffer solution	Buffer capacity	Dilution value
0·05 M potassium tetraoxalate	0·070	+0·186
Sat. (at 25^{oC}) potassium hydrogen tartrate	0·027	+0·049
0·05 M potassium hydrogenphthalate	0·016	+0·052
0·05 M piperazine phosphate	0·037	negligible
0·025 M potassium dihydrogen phosphate 0·025 M disodium hydrogen phosphate	0·029	+0·080
0·01 M borax	0·020	+0·010
Sat. (at 25^{oC}) calcium hydroxide	0·090	−0·280

* from Mattock[1] (pH measurement and titration; Heywood & Co Ltd)

2.7. Colorimetric methods of pH measurement

Details of colorimetric measurement of pH are available in various books and reviews. Since we are basically concerned with electrometric methods we shall only consider indicator methods very briefly.

Indicators are usually dye molecules. It is conventional to assume that indicators behave as weak acids or bases and that the degree of dissociation into, say, the anion form and a hydrogen ion is governed by the hydrogen ion concentration in solution. The indicator molecule and its anion give different colours in solution and the resultant colour intensity is therefore proportional to the degree of dissociation and hence the hydrogen ion concentration in solution. This theory gives an over-simplified picture of indicator action but suffices for most experimental explanations, and the acid dissociation constants of indicators are useful for purposes of comparison.

Indicators are now used only where rapid and approximate pH determinations are required. And even in such situations their main advantage is their low cost since electrometric measurements are equally quick and far more accurate, accuracy with visual techniques being less than $\pm 0 \cdot 1$ pH units.

Approximate pH indication

A rough check of the pH of a solution can be obtained by using a mixture of indicators. The 'Universal' indicator solutions which are such indicator mixtures cover a wide range of pH values. Thus the BDH Universal indicator ranges from $3 \cdot 0$ pH to $11 \cdot 0$ pH, each level being characterised by a particular colour. Commercially available indicator papers use the same principle.

Various more refined and accurate indicator methods are available. Most of these involve comparison, e.g., with graded buffer solutions containing the same amount of indicator (as in the BDH 'small comparator' and BDH 'Capillator') or with coloured glasses which have been standardized against known buffer indicator solutions as in the Lovibond comparator Tintometer).
However, to achieve any degree of accuracy, the estimation of colour or light intensity must be instrumental rather than visual. Photometric techniques have been used by some workers but they have various disadvantages, extra time and expense involved being the major ones.

In laboratory titration experiments the situation is different: indicators play a very important role and until recently potentiometric titrators were only used in special cases. The large number of reliable direct reading pH meters now available, however, is changing this state of affairs.

3. MEASUREMENT OF pH IN CHEMICAL LABORATORIES

3.1. General principles

Desirable characteristics of a pH determination and detailed procedures employed will obviously depend on the particular problem. The type of material, its stability and buffer capacity and the accuracy required all have to be considered.

Some of the general principles which guide the determinations of pH in chemical laboratories (see Bates[2]) are summarised below.

1. The pH of the unknown or test solution is obtained by comparing its pH with that of a standard solution at the same temperature.

2. The accuracy of the measuring device should be tested by determining the pH of at least two standard solutions. The two standards should bracket the pH of the unknowns.

3. Errors caused by fluctuations in temperature and by the residual liquid junction potential are minimised by standardising the assembly at a pH value close to that of the unknown solution.

4. Care should be taken to select the glass electrode most suitable for the pH range of the unknown solutions.

3.2. The standard pH method

Let us now consider the standard pH method described by the American Society of Testing Materials.[3]

Standardization of the assembly

The instrument should be allowed to warm up thoroughly and the amplifier brought into electrical balance in accordance with the manufacturer's instructions. The electrodes and sample cup must be washed three times with distilled water and dried gently with absorbent tissue. If the salt bridge is not of the continuous flow type, a fresh liquid junction should be formed, and the temperature dial of the meter adjusted to the temperature of the unknown solution.

Two standard buffer solutions, their pHs preferably above and below that of the unknown solution, should be brought within 2° of the temperature of the unknown. Then, the electrodes are immersed in a portion of the first standard solution. The range switch should be switched to the required range, and the standard-izing or asymmetry potential knob turned until balance is obtained at the pH of the standard (obtained for the operating temperature from Appendix 2 and 3). The process is then repeated with additional portions of the first standard until the instrument remains balanced ($\pm \cdot 02$ pH) for two succeeding portions without adjustment. Care must be taken that the temperature of the electrodes is the same as that of the solution. The temperature of the electrodes wash water and standard solutions should be as close as possible to the temperature of the unknown solutions.

The electrodes and sample cup are then again washed three times and a fresh liquid junction formed. The second standard is then placed in the cup, the instrument brought to balance and the pH read without changing the position of the asymmetry knob. Readings are repeated with additional portions of the second standard till successive readings differ by less than $\pm$ 0.02*pH* units. If the assembly is operating satisfactorily then the reading obtained for the second standard should come within $\pm 0 \cdot 02$ units of its assigned pH. The instrument should always be restandardized after a period of disuse during which the amplifier has been turned off and a final check should also be made at the conclusion of a series of measurements. Apart from these checks, occasional restandardization may be necessary during a set of determinations.

If the instrument is being used to measure the pH*(s) of alcohol-water solvents, there may be some difficulty in reading values directly as the zero adjustment provided may be insufficient to compensate for the large changes of EMF of the cell. If this is the case, the instrument may be adjusted with the standard solution of known pH*(s) to a convenient arbitrary point on the scale. A measured difference of pH* is then determined and added to the pH*(s) value of the standard.

pH of test solutions

The electrodes and sample cup should be washed thoroughly and dried with absorbent tissue. The sample cup is then filled with test solution, a fresh liquid junction formed and a preliminary pH value determined. Additional portions of test solution are then used till successive readings differ by less than $\pm 0 \cdot 02$ units (for well buffered solutions) or $\pm 0 \cdot 1$ units (for water or poorly buffered solutions).

If a glass electrode is used and an accuracy of greater than $0 \cdot 01$ units is required special attention must be paid to 1. temperature control, 2. electromotive efficiency of the glass electrode, 3. variable liquid junction potential. When the unknown and standard solutions contain different types of ion and have different ionic strengths inconsistent results are often obtained; this can usually be attributed to residual liquid junction potential. The temperature of measurement also influences accuracy and reproducibility; ideally it should not differ appreciably from ambient temperature. In general the pH of alkaline buffer solutions is more sensitive to temperature changes than that of acid solutions. Sensitivity to temperature fluctuations, glass electrode errors and selectivity and the changing liquid junction potential thus make measurements at high pH much less accurate than those in the region near neutrality.

If the solution to be measured is only slightly buffered, the procedure recommended is similar to that previously described but with vigorous agitation. Six or more portions may have to be used to obtain a reading that drifts less than 0·1 units in 2 minutes especially when the electrodes have been immersed in a buffer solution immediately prior to the measurement.

3.3. Design of pH meters

Having described the standard pH method we can now consider the construction of some common types of commercially available pH meters. The basic object of these instruments is the accurate determination of the EMF of a cell of extraordinarily high resistance. Criteria important in their design have been listed by Clark and Perley.[4] They may be summarized as follows:

1. Measurements must be unaffected by variations in the resistance of the glass electrode and the reference electrode during experiments.
2. Measurements must be unaffected by grounding or ungrounding the solution which is being measured.
3. The indicator should be AC powered and its operation should be independent of changes in power supply within the range 100 to 130 v.
4. Temperature compensation should be possible and preferably automatic.
5. Changing vacuum tubes should not affect circuit constants.
6. It should be possible to connect a recorder to the measuring system with no loss of precision. A single pH range from 0–14 would be desirable for recorder operation.
7. The instrument should be adapted to a wide range of EMF

measurements in high and low impedence systems.

Clark & Perley also give a detailed analysis of the four main types of electrical layouts used in pH meters; these will be now considered briefly.

Method 1

This is one of the earliest methods used; the aim here as in the other three methods is to measure the voltage originating in various types of electrodes without drawing appreciable current. A null type circuit is used, the detecting element being made up of one or more vacuum tubes; the input voltage from the electrode system is balanced as nearly as possible by voltage from a source such as a manually operated potentiometer. The maximum current by such a circuit can be made very small by using sufficient amplification; the correct grid bias voltage on the input tube also decreases the grid current. Theoretically the electrode voltage may be exactly balanced and electrode voltage or pH values are given by potentiometer readings.

This system has various defects; special precautions have to be taken to reduce grid current errors, and a very stable direct voltage supply is essential for both the electron tube and balancing voltage in order to reduce zero error drift.

Method 2

This method uses a direct-coupled DC amplifier to develop the balancing voltage which in this case is a feed-back voltage proportional to the difference between balancing feed-back and input voltage from the electrodes. If sufficient gain is used, current drawn from the electrode system is negligible and electrode voltage can be measured by metering the feed-back current, see figure 5. When the gain is adequate the range of the instrument is almost unaffected by change in the power supply and normal changes in characteristics of the tubes; however the zero readings may alter considerably. Temperature compensation can be conveniently arranged by making the feed-back current flow through a temperature sensitive resistor. The value of the balancing voltage and hence the range of the instrument is dependent on the resistance of the resistor and this is a function of its temperature. Placing the resistor in the solution to be measured thus provides automatic temperature compensation.

Instruments using this method need precautions to minimize grid current error. They also require a very stable direct voltage

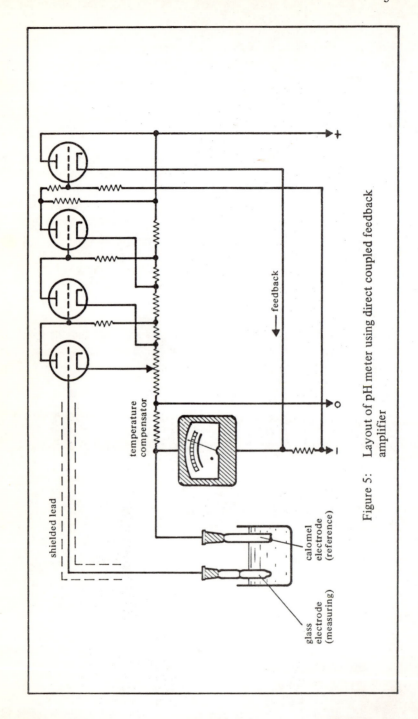

Figure 5: Layout of pH meter using direct coupled feedback amplifier

37

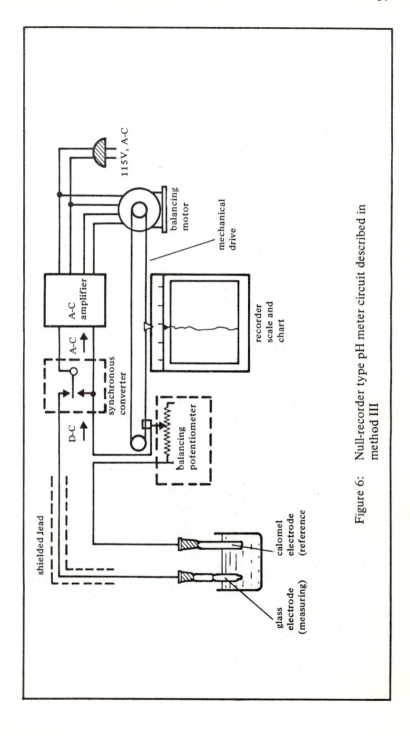

Figure 6: Null-recorder type pH meter circuit described in method III

supply for the direct-coupled DC amplifier to reduce zero-drift error. The ageing of tubes or changing them makes circuit readjustment necessary to correct for changes in zero. Precision of measurement also depends on the precision of calibration of the indicating meter.

Method 3

In this method, figure 6, a synchronous convertor chops the voltage difference between the electrodes and feed-back into alternating voltage; the signal is then amplified and used to adjust the feedback voltage electromechanically and obtain the null point. The electromechanical arrangement is usually (e.g., in the L & N Speedomax type) a motor operated slide wire from which the feed-back voltage is obtained.

The input circuit is isolated from the first stage of amplification by a capacitor; the precision of measurement is, therefore, unaffected by grid current errors. Also, since the only source of voltage in the input circuit, apart from the electrode system, is the balancing potentiometer and since this is standardized periodically against a standard cell, zero drift is negligible. Changes in tube characteristics affect only the gain of the instrument. More than adequate gain is provided, therefore this introduces no error.

Method 4

Chopper type feed-back stabilized method

This method combines features of Methods 2 and 3 and falls between them in instrument cost. It is essentially the same as Method 2 except that the difference voltage is converted to AC for amplification then back to DC and the circuit completed as in Method 2.

The direct difference voltage is chopped and amplified as in Method 3. The resulting AC voltage is rectified by a phase-sensitive circuit and fed back into the input circuit as the 'balancing' voltage in opposition to the 'direct' voltage. The input voltage is indicated by a measurement of the feed-back current as in Method 2. As in 2 and 3, the gain is sufficient to keep the range of the instrument unaffected by mains voltage fluctuations or tube characteristic variation and to make the current from the electrodes negligible. Since the only voltage or current sources in the input circuit are the electrode and balancing voltages, the zero will be stable and independent of mains and tube variations. The input circuit is protected from grid currents by blocking capacitors.

39

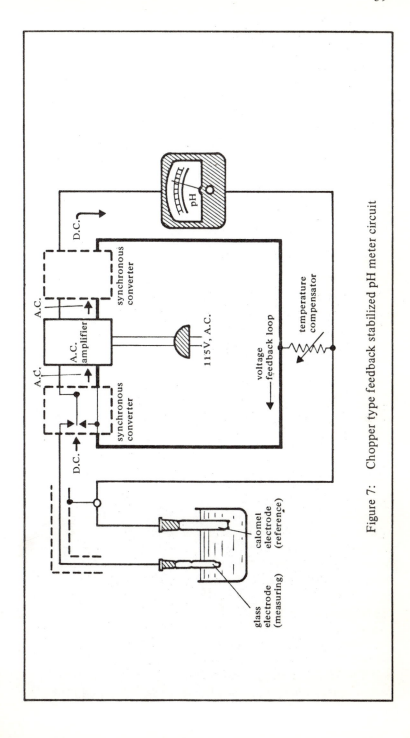

Figure 7: Chopper type feedback stabilized pH meter circuit

Temperature compensation is the same as in Method 2.

The precision with which the meter indicating the feedback current is calibrated determines the precision of measurement. It need cause no appreciable error (e.g., L & N stabilized pH indicator).

3.4. Errors of pH meters and performance tests

Table 2 shows the most important instrumental errors in percentages of scale reading as summarized by Perley. Since the pH meter is made to read correctly at the pH of the standard, these percentage errors apply strictly only to the difference of scale reading between standard and unknown.

Table 2
Errors of pH meters

Cause of error	Max limit error % scale reading
1. Calibration and adjustment of temperature compensator	0·10
2. Zero adjustment (for potentiometric type instruments)	0·10
3. Calibration of slide wire	0·30
4. Adjustment of standardizing current	0·10
5. EMF of standard cell (direct reading type instrument)	0·05
6. Calibration of deflection meter	1·00

Before using a new pH meter, the millivolt scale should be calibrated against a known EMF source and the temperature compensator accuracy should be estimated. We shall summarize now these two performance checks as described by Bates.[2] For work where high accuracy is required these tests must be repeated at least once a year.

The calibration of the EMF scale is done by comparison with a known potential supplied by a potentiometer accurate to 0·1 mV. The known EMF is applied directly to the meter and again through a resistor R rated at 100 to 200 megohms, the resistor being mounted in a metal shield to avoid capacity pick-up.

The negative EMF terminal of the potentiometer is connected through the series resistor to the glass electrode terminal of the meter and the positive terminal of the potentiometer is connected to the calomel electrode terminal. The potentiometer is then brought to electrical balance in accordance with the manufacturer's instructions, the proper EMF range selected and starting with a value of zero the applied potential is increased in steps of 100 mV. The readings of the meter at balance are noted for each value. An error of up to 1 mV per 100 mV increment of applied voltage is usually permissible, but the cumulative error at the upper end of the scale should not exceed ± 5 mV.

The second important performance test is a check on the reproducibility of the EMF calibration curve at various settings of the manual temperature compensator. The meter is brought to balance and switched to the pH scale. The potentiometer connection is then made by means of the meter switch or push button and the asymmetry potential knob is rotated till the meter is balanced at the pH of zero EMF for the instrument. The applied potential is then increased in increments of $(RT \ln 10)/F$ volts at the temperature setting of the temperature compensator, the pH dial readings noted and pH error computed. The increments of voltage correspond to increases of exactly one pH unit at the respective temperature, for example, 59·16 mV at 25°, 56·18 mV at 10°, etc. If zero impressed EMF corresponds to a pH of 7, the above procedure allows calibrations of half the scale range the other half can be studied by reversing the connections of the reference potentiometer. Bates suggests the following convenient table for the results of this calibration.

Table 3
Suggested table for calibration data
$t = 25°$ meter no.

Potentiometer voltage	Theoretical pH Range: 0 to 7 pH	Scale reading	Error
0	7·00		
0·0592	6·00		
0·1183	5·00		
0·1775	4·00		
0·2366	3·00		

Potentiometer voltage	Theoretical pH Range: 0 to 7 pH	Scale reading	Error
0·2958	2·00		
0·3549	1·00		
0·4141	0·00		

	Range: 7 to 14 pH		
0	7·00		
0·0592	8·00		
0·1183	9·00		
0·1775	10·00		
0·2366	11·00		
0·2958	12·00		
0·3549	13·00		
0·4141	14·00		

3.5. Automatic titrators

Automatic titrators consist essentially of three parts, a pH meter, an amplifier control unit and an automatic burette valve. If the pH at the end point of a titration is known, the automatic titrator can conduct the entire experiment and, when used in conjunction with recording equipment, leaves a permanent record of the titration curve. The accuracy is usually about 0·1% if the pH change at the endpoint is sharp, and time taken for one titration is less than three minutes. The operator fills the burette and sets the dial to the desired endpoint pH. Raising the vessel holder into position, he initiates the titration and energizes the stirrer. When the endpoint is reached, the stirrer is turned off automatically and a panel light signals the conclusion of the titration.

The electrode assembly of the titration cell should be standardized with a standard buffer solution although the accuracy of this standardization is often of secondary importance in view of the large pH changes at the endpoint particularly in some neutralization titrations.

In some commercial instruments the reagent is added intermittently, each addition diminishing in duration as the endpoint is approached, others use a reduced flow rate near the endpoint.

pH stats

The pH stat is a modification of the automatic titrator; it records the amount of reagent required to maintain a constant preset pH as a function of time. The control point of a pH stat is analogous to the endpoint of an automatic titrator, when it is reached, however, the instrument does not turn itself off. Instead, it continues to operate, neutralizing the acid or alkali produced in the system. Titrations at constant pH are often necessary in enzyme systems. Jacobsen, Leonis, Linderstrøm-Lang and Ottesen (*Methods of Biochemical Analysis*, Vol.4, p.171, Interscience Publishers 1957) have reviewed some uses of the pH stat. Of commercially available instruments, two important ones are the Radiometer Titrator and Titrograph (which can function as an automatic titrator as well as a *pH* stat) and the Polarad Titrator AT-2A.

3.6. Differential titrators

A differential titrator eliminates the need for curve plotting in a potentiometric titration. Here, the reference electrode is made exactly similar to the indicator electrode. It dips into a solution buffered to the end point pH and a liquid junction connects it to the solution being titrated. In the cell thus formed, pH values and hence EMF contributions of the sample and reference solutions become identical at the end-point and the resultant EMF tends to zero.

Quinhydrone electrodes are commonly used in differential titrators, glass electrodes being unsuitable because of their high resistance. Various refinements and modifications of this basic differential titration technique have been developed and highly accurate differential titrators are commercially available. The technique has also been applied to non-aqueous solutions, platinum and chloranil indicator electrodes being used.

4. INDUSTRIAL pH MEASUREMENT AND CONTROL

4.1. Applications of pH in industry

pH measurement in industry has been classified into four main systems:
1. Preparations where efficiency is dependent on the pH of the solutions involved, e.g., penicillin manufacture.
2. Quality control: situations where uniformity of product depends on pH control at some stage, e.g., the paper industry.
3. Neutralization of effluents.
4. Corrosion inhibition, e.g. in modern high pressure boilers

4.2. pH control cells

There are two main types of pH control cell used in various control installations: the *immersion* or *dip unit* and the *flow unit*. The first of these in a common form consists of a hood of metal, plastic or hard rubber containing glass and calonel electrodes and a resistance thermometer mounted by means of threaded connections and made water-tight by sealed washers. The thermometer automatically corrects for the change of *dpH/dE* with temperature. The hood is supported at one end by a metal pipe which also shields the electrode lead wires, the element being protected by bar guards. This type of unit is designed for use under low external pressures but it usually also works satisfactorily at depths of 60 ft or more. Another common type of dip unit (manufactured by Beckman Instruments Inc) is the electrode gland assembly. This consists of three threaded glands in which are mounted the glass electrode, the reference electrode and the temperature compensator sensing element, (figure 8).

While the dip unit is designed for direct immersion in a tank or vat, the flow unit is meant for use in a channel through which the process solution flows; it thus measures the pH of a continuous sample.

In this case the electrodes and resistance thermometer are fixed in a mounting plate which is bolted to a metal, glass or plastic flow chamber. The rate of flow through the chamber is unimportant in most applications as long as the liquid sample in

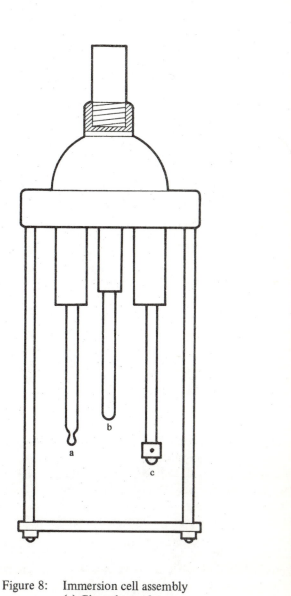

Figure 8:   Immersion cell assembly
           (a) Glass electrode
           (b) Resistance thermometer
           (c) Calomel reference electrode

the chamber is representative of the solution in the main tank. However, a rapid flow may be required in some cases to avoid deposition of fine particles on the electrodes, and large fluctuations in the rate of flow must be avoided as they disturb liquid junctions.

4.3. Some common problems

Industrial installations operating at high or low pressures present the commonest problems. Of these the most difficult to overcome is the establishment of a stable liquid junction. Calomel electrodes with junctions of fibre, sleeve, palladium annulus, porous sponge and capillary types are available commerically but none is entirely satisfactory, especially at high pressures. One of the most effective means of obtaining free-flowing junctions at high pressures uses an external source of air pressure. Beckman Instruments manufacture pressurized glass electrodes and Polymetron A.G. (Zurich) have a differential pressure regulator which keeps the reference electrode and salt bridge at a pressure of 0·1 atmospheres above the pressure of the process stream.

At temperatures above 80°C, calomel slowly disproportionates and calomel electrodes become more and more inaccurate; silver–silver chloride electrodes can be substituted up to 100°C and thallium amalgam – thallium (I) chloride electrodes (manufactured by Jena Glaswerk, Mainz) work satisfactorily up to 135°C. These and various other kinds of glass electrodes for use at high temperatures are available commercially. Processes involving sterilization of plant components present a special problem since electrodes are subjected to temperatures of 120–130°C and then used to measure at 25°C without loss of prior buffer calibration. A glass electrode which has to undergo sterilization has to have a suitably resistant membrane. Several commercial electrodes are available which can stand a large number of sterilizations before exchange deterioration occurs. In a successful control system too the inner reference electrode must return to its original potential after the sterilization cycle and the total asymmetry potential must be unaffected. Thermal cycling of the calomel reference electrode is not advisable and can be avoided where the electrode is connected to the system through a remote liquid junction having bacteria proof properties. In many applications the sterilization can be eliminated by the use of ceramic bacterial filters. Specially designed steam-sterilizable electrodes are also available. [5]

In many industrial installations it is necessary to provide some arrangements for self-cleaning of electrodes. For example, in paper treatment a high rate of flow is employed and steam is aimed directly at the glass electrode. The mild abrasive action of paper pulp maintains a clean glass surface. Decomposition on electrodes can also be reduced by the presence of a small amount of detergent or by periodic automatic flushing of the assembly with dilute acid. Automatic wiping arrangements are also available commercially, manufactured by W. Ingold (Zurich).

4.4. Principles of automatic control

A detailed discussion of methods of automatic pH control is beyond the scope of this book. We shall, however, consider here very briefly the working of a general automatic control system. Three of the most important characteristics of such a system are *capacity*, *transfer lag* and *dead time*. Capacity represents the ability of the system to absorb the control agent without change in the process variable. Transfer lag results from the inability of the system to supply the required amount of corrective agent instantly on demand. Dead time is simply time delay in any part of the system; measuring lag and controller lag are a part of dead time. Modern controllers have four basic control patterns or modes of compensatory response to pH or other variables. These are 1. two position control, 2. proportional position control, 3. reset responses, 4. rate action. 3. and 4. are usually employed in conjunction with 2.

Two position regulation is the simplest kind of control arrangement, the reagent valve is always in one of two positions, the first causing a flow of corrective reagent greater than needed and the second a flow less than needed. On–off and high–low controls are examples of this.

Proportional position control uses a sliding-stem control valve, in which the position of the valve stem (which is continuously variable) is proportional to the value of the controlled variable at all times. The proportionality factor is dependent on the adjustment of the proportional band of the controller. The proportional band, or throttling range, is defined as the range of values of the controlled variable that corresponds to a full range of the final control valve, and it is usually expressed as a percentage of the full scale range of the controller. Widening the throttling range lowers sensitivity and although narrowing the range diminishes the magnitude of deviations, the oscillations following a sudden change are increased. Proportional control initiates corrective action immediately a deviation is apparent and obviously provides more accurate control than two-position regulation.

To provide stability under a variety of load conditions it is often necessary to superimpose 'reset response' on the proportional control pattern. The resultant is called *proportional-reset control* and it causes an advance in the valve position determined by the magnitude and length of time of the deviation in the process variable. Reset action also brings the variable back to the set point even though a new valve position is necessary to meet a sustained change of load.

The fourth control mode, 'rate action', is usually applied to processes with a large dead time or transfer lag. In pH control, rate action supplies an added valve adjustment determined by the rate at which the pH is deviating from the set point; this initial large corrective action being intended to counteract the effect of process lag. Obviously rate action acting in reverse can also prevent overshoot as pH returns to the balance point.

4.5. pH control and buffer action

pH control is nearly always nonlinear. There are two main reasons for this: first that pH is a logarithmic function of concentration and the second that buffer capacity varies extremely rapidly and non-linearly with pH. Wide band proportioning action in conjunction with reset response is the most frequently used mode of control. Its effectiveness depends largely on the rate of change of pH with volume of corrective reagent near the control point. If this rate is large (for example, near the end point of the neutralization of a strong alkali with a strong acid), proper control is very difficult. The problem may be solved by using a weak acid or base as the control agent or by increasing the capacity of the mixing chamber. However, when pH is insensitive to addition of corrective agent, accurate control may once again be difficult. Direct measurement of pH of solutions of strong acids or bases, for example, can be used for pH control only near the neutral point.

4.6. Control elements

Selection of the best means of regulating the corrective agent is most important; sliding stem valves and dry chemical feeders are two common control elements both operated by electrical or pneumatic controllers.

The sliding stem valve, pneumatically controlled, is quite

simple to set up and fairly precise. It offers a wide range of throttling controls and is available in a variety of inner structures. However, if the corrective agent is a slurry it tends to get blocked and it is then safer to use a Saunders patent valve. This throttles the flow by pressing a flexible diaphragm towards a raised seat in the valve shell.

Feeders for solid reagents eliminate the need for large solution tanks and for corrosion resistant valves, fittings and pipes. There are various types of feeders available, particularly for solids; belts and vibrators are most common but bin valves, and constant weight feeders are also used. Rate of delivery can be controlled by electrical or pneumatic proportional action.

5. MEASUREMENT OF pH IN BIOLOGICAL AND MEDICAL LABORATORIES

Acid-base balance has considerable influence on the behaviour of biological systems. pH values which usually vary within fairly narrow limits are an important source of information. In this chapter we shall discuss briefly the techniques for measuring pH in biological systems and the problems encountered.

5.1. Blood pH

A number of chemical equilibria contribute to the maintenance of physiological neutrality in blood. The one most often investigated by pH measurement is the bicarbonate-dissolved carbon-dioxide buffer system. The pH range covered by this is usually 7·2 to 7·5 and the Henderson–Hasselbalch equation, see equation 36., for this system is

$$pH = pK_{HA}^a + \log \frac{C_{HCO_3^-}}{C_{(H_2CO_3 + CO_2)}} \qquad \ldots 46$$

where $C_{HCO_3^-}$ is the concentration of bicarbonate ions
$C_{(H_2CO_3 + CO_2)2}$ is the concentration of dissolved carbon-dioxide
K_{HA}^a is the apparent first dissociation constant of carbonic acid
Writing the equation in terms of the partial pressures of the gaseous carbon-dioxide of the system we get

$$pH = pK_{HA}^a + \log \left(\frac{tCO_2 - \alpha pCO_2}{\alpha pCO_2} \right) \qquad \ldots 47$$

where α is the solubility coefficient of carbon-dioxide in blood.

Determination of (tCO_2) (by titration) enables evaluation of pCO_2 if pK_{HA}^a and α are known. Obviously the logarithmic relationship between pH and pCO_2 expressed in equation 47. makes accurate pH measurement particularly important.

pK_{HA}^a and α values have been assumed to be constant by various workers. In fact pK_{HA}^a shows wide variations with temperatures and pH values. Detailed values of pK_{HA}^a of normal blood are available in a nomogram by Severinghaus et. al.[6] However, the following procedure eliminates use of pK_{HA}^a values. First measure the pH of a sample of the whole blood. Separate another sample and equilibrate each of the two portions of plasma with known partial pressures of carbon-dioxide and determine the pH in each case. If pH varies linearly with log $1/pCO_2$ for all three measurements, the unknown pCO_2 for the sample representing the whole blood can be obtained by interpolation.

This assumes no variation in pK_{AH}^a and α between the whole blood and plasma. A more recent technique uses whole blood throughout[7].

Temperature control is particularly important in the measurement of blood pH, temperature correction being impossible for various reasons. Temperature coefficients of blood pH may change with individual samples and separation of plasma at a temperature even slightly different from the temperature of measurement may cause inaccuracies. Maintenance of constant temperature is therefore essential and storage of both glass and reference electrodes at the temperature of measurement is advisable. The blood should be stored at low temperature (about $2°C$) to avoid glycolysis (the slow formation of acids) and anaerobic conditions of measurement should be maintained.

Various types of electrode systems have been used in the measurement of blood pH. Figure 9(a) shows a commercial form of the 'Stadie System'.

Blood is admitted to the jacketed tube in which the glass electrode is inserted. Contact with the reference electrode is made by turning the stopcock through $90°$.

In another electrode system the glass electrode acts as the plunger of a syringe, the liquid junction is established at the opening of the syringe barrel and the male luer tip fits into a female socket. The 'dead-space' between the electrode bulb and the inner surface of the syringe barrel is usually filled with dilute heparin solution, before blood withdrawal; this prevents coagulation and carbon-dioxide loss.

Figure 9(b) shows another commonly used electrode system in which the liquid junction is inside a fine capillary, the lower (salt-bridge) solution being stopcock-controlled and the cell chamber situated just above the capillary. Mattock[1,8] has reviewed some representative electrode systems for measuring the pH of blood.

In all the systems mentioned it is important to check glass electrodes for poisoning. This can be done by checking pH response against buffer standards and is particularly important in apparatus where blood remains in contact with the electrode for long periods (e.g., 30 to 40 minutes). The poisoning appears to be

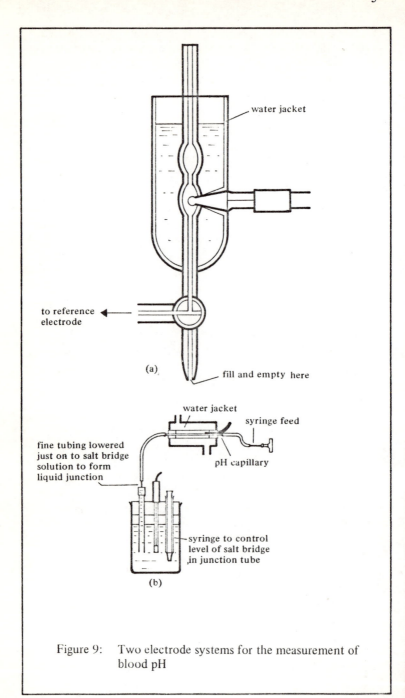

water jacket

to reference electrode

(a)

fill and empty here

water jacket

syringe feed

fine tubing lowered just on to salt bridge solution to form liquid junction

pH capillary

syringe to control level of salt bridge in junction tube

(b)

Figure 9: Two electrode systems for the measurement of blood pH

caused by protein decomposition and can sometimes be removed by wiping the electrode carefully after standing it in pepsin and O.IN hydrochloric acid. In cases where the restricted flow liquid junction is a part of the apparatus, electrodes should be stored in isotonic saline. Isotonic sodium chloride salt bridges can be used instead of saturated potassium chloride ones. These reduce coagulation at the interface and the resulting clogging of the liquid junction. However, recent investigations have shown that the concentration of salt in the salt bridge effects the observed pH of blood or plasma.

Two different solutions have been proposed as standards for blood pH measurements. One, established by the National Bureau of Standards, has an ionic strength of 0·1 and contains potassium dihydrogen phosphate and disodium hydrogen phosphate in the molal ratio 1 : 3·5. The other described by Semple, Mattock and Uncles,[9] contains the same salts in the molal ratio 1 : 4 (i.e., 0·01 M (KH_2PO_4) and 0·04 M (Na_2HPO_4) with an ionic strength 0·13. The first solution has been found to give a pH of 7·388 at 38°C, the second one gives a pH of 7·416 ± 0·004 at 37·5°C and 38°C.

The most widely used electrode in blood pH measurement and also for other clinical applications is that of Astrup-Radiometer. It is shown in figure 10. It requires 20 to 25μl of capillary blood for a determination. Measurement against a constant buffer gives a standard deviation of ± .001pH

5.2. pH of gastric fluids

The pH values to be measured in this case lie between 1 and 9 so accuracy need not be very high. Both sampling techniques (involving withdrawal of a small amount of the gastric juices from the stomach followed by a laboratory pH measurement) and various arrangements for measurement within the body have been tried. The electrodes used are mainly glass and antimony and the following are some of the problems encountered.

1. The position of the calomel reference electrode appears to affect pH values. If the salt bridge connection is made at some convenient point outside the stomach, for example, potentials such as that of the skin may interfere with measurements. However, recent work suggests that no error arises from siting the liquid junction in the mouth.

2. Stomach mucus may accumulate on the electrode, slowing its response.

3. The electrode should not be in contact with the stomach lining

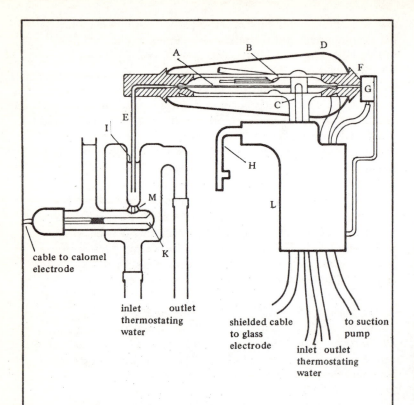

cable to calomel
electrode

inlet   outlet
thermostating
water

shielded cable
to glass
electrode

to suction
pump

inlet outlet
thermostating
water

A is a pH sensitive glass capillary tube surrounded in
buffer in tube B. The electrode, C, and buffer fluid
are connected to the pH meter by an electric cable in
L, the shaft of the electrode. Thermostated water
circulates in D. The short polyethylene tube F
provides an extension for the capillary A. A is
filled with the sample through the polyethylene tip E
using the suction device G. K is the thermostated
reference calomel electrode connected through the
disc M to the salt bridge I. The liquid junction is
formed between the saturated KC1 in I and the
sample in E. By means of a metal rod H, the
electrode may be placed in a special holder in the
measuring or resting position.

Figure 10: Thermostated capillary glass electrode for blood pH

as this produces suprious EMF effects.

4. If antimony electrodes are used, complex-forming acids and poisoning metals should be avoided.

A stomach glass electrode must be small in diameter and length and have a low electrical resistance and mechanically secure insulation. Often it is constructed from screened cable, a 2·5 mm diameter cable giving an electrode 3 mm in diameter, the rigid length being about 6 mm. The inner reference electrode can be made by silvering the inner surface of the glass. Antimony electrodes are simpler to make, insulation being easier. Antimony is cast in a capillary tube and a length about 5 mm soldered to a PVC coated multiwire lead, the joint carefully filed smooth and covered with a protective coating of 'Araldite'. As a final protective layer, silicone rubber sleeving is cemented over the joint with silicone resin.

A convenient method of measuring pH of gastric juices has been described by B. Jacobsen and R. S. Mackay[10]. It involves the use of a '*pH* endoradio-sonde'. This consists of a thin membrane of a particular polymer (20 per cent polyacrylic acid and 80 per cent polyvinyl alcohol) which expands and contracts reversibly with changes in pH. The volume change of the membrane can be transferred to a transducer which forms the inductance of a transistor oscillator. Thus the volume change causes a variation in transmitted frequency which can be picked up by a loop antenna of a radio receiver. This technique has now been applied in conjunction with a glass electrode unit.

5.3. pH of dental plaque

The range of pH values in dental cavities is similar to that in the stomach. Size of electrodes rather than accuracy is the main problem. Various micro-electrodes are available commercially, most of them in glass or antimony and of the curved 'spear-tip' shape.

Measurements made with antimony micro electrodes have been found to differ from those made with glass micro electrodes by 0·3 to 1 pH unit; possibly intra-oral fluids affect the antimony response.

A glass micro electrode has been described by Charlton[11]. It consists of a MacInnes–Dole glass bulb about 0·25 mm in diameter attached to a lead-glass capillary stem mounted in a polythene tube· The inner electrode consists of a platinum wire dipping into 0·1 M hydrochloric acid solution containing quinhydrone.

5.4. pH measurement in tissues

Various techniques have been used for these measurements. However only glass electrode methods appear to give reliable results. Dole[12] has reviewed work in this field up to 1940.

Tissue is a very general term and electrodes employed for different kinds of tissue vary considerably. A glass micro electrode somewhat similar to Charlton's dental probe has been used to study intercellular pH changes in crab muscles. While brain pH measurements are carried out with microspear electrodes.

The main difficulty with tissue pH measurements is that electrodes often affect tissues, damaging them and causing a change in their pH. Skin pH measurements have become increasingly important. Flat headed glass electrodes are commonly used in these experiments and the reference electrode junction is placed as close as possible to the glass. The pH-sensitive part of the electrode must be thoroughly wetted by the fluid whose pH is desired. Ceramic or fibre liquid junctions establish the best reference contact with the skin.

6. REFERENCE AND MEASURING ELECTRODES

The basic equipment for measuring pH consists of (a) an electrical
circuit for measuring EMF, (b) a galvanic cell consisting of
1. a reference electrode, including solution R (with which it is in
contact);
2. a salt bridge and
3. a measuring electrode and a vessel to hold the solution unknown
or standard in which it is immersed. The cell can be represented
by:

Measuring electrode (an electrode reversible to the hydrogen ion)	Soln. X or S	Salt bridge	Soln. R	Reference electrode
				... 48

The reference electrode is usually a calomel electrode. The measuring electrode can be any one of several electrodes reversible to
hydrogen ions; of these, the hydrogen electrode is the ultimate
standard for determination of pH values, but it is not always
the most convenient,being associated with various experimental
difficulties. Other electrodes reversible to the hydrogen ion are
employed for routine pH measurements, e.g., the glass electrode
or the quinhydrone or antimony electrodes.

6.1. The hydrogen electrode

The hydrogen electrode is formed by bubbling pure hydrogen
gas over a wire or small foil placed in the experimental solution.
The wire or foil is of platinum coated with finely divided platinum
black, platinum black being a catalyst in establishing the
equilibrium

$$H_2 \text{ (gas)} \leftrightharpoons 2H^+ + 2e \qquad \qquad \text{...} 49$$

The electron charge is taken up by the platinum till the potential
between the electrode and solution balances the tendency of the
hydrogen molecules to ionize. In the standard hydrogen electrode
the hydrogen gas is at 760 mm pressure and the solution of
hydrogen ions is of unit activity. The potential of the standard
hydrogen electrode has been chosen as zero. It is commonly used

in conjunction with a calomel reference electrode.

The hydrogen electrode has various drawbacks the most serious being that platinum black is 'poisoned' by a variety of compounds (e.g., the cyanide ion, hydrogen sulphide, arsenic compounds, cations of metals more noble than H_2, etc.). These poisons inhibit the reversibility of the electrode process (equation 49) by being preferentially adsorbed on the electrode. The high reactivity of hydrogen in contact with platinum is also a disadvantage since it requires the exclusion of easily reduced aromatic compounds from the experimental solution.

In potassium hydrogen phthalate solutions where the platinum black electrode is unsuitable because of its reducing action, palladium coated electrodes have been found to give constant and reproducible potentials. Details of various kinds of hydrogen electrodes and their construction have been discussed by Bates.[2]

6.2. The glass electrode

In its most common form the glass electrode consists of a thin glass bulb inside which is mounted a reference electrode immersed in a solution of constant pH containing the ion to which the inner reference electrode is reversible. The MacInnes–Dole electrode (figure 11) for example, is formed by fusing a membrane of pH sensitive glass across the end of a glass tube. The inner cell consists of a silver chloride or calomel electrode in hydrochloric acid or a buffered chloride solution. The type of the inner electrode and inner solution determines both the potential with respect to an external reference and the temperature coefficient of the potential. The variation of potential of the glass electrode with pH can be measured with a cell of the type

$(Ag)|AgCl|O\cdot IN\ HCl\|glass\|Test\ solution|Saturated\ KCl|calomel$

$$\ldots 50$$

the glass membrane being represented by double lines. If E_1 and E_2 are the values of EMF for this cell in test solutions of pH_1 and pH_2 respectively, the pH response R_{pH} (in millivolts per pH unit) is given by

$$R_{pH} = \frac{E_2 - E_1}{pH_2 - pH_1} \qquad \ldots 51$$

An ideal glass electrode would react to changes in hydrogen ion activity in exactly the same manner as the standard hydrogen gas

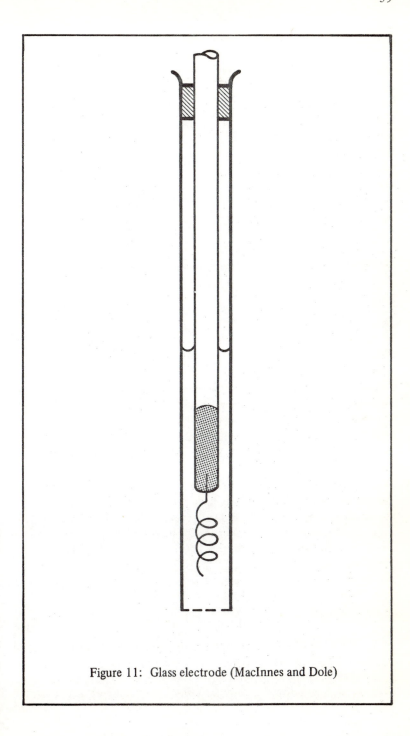

Figure 11: Glass electrode (MacInnes and Dole)

electrode and we would get

$$E_2 - E_1 = \frac{RT \ln 10}{F} (pH_2 - pH_1) \qquad \ldots 52$$

and thus ideal pH response would be ($RT/F \ln 10$) or $2 \cdot 3036 RT/F$ volts per pH unit, i.e.,

$0 \cdot 05420$ V at $0°C$

$0 \cdot 05916$ V at $25°C$

and

$0 \cdot 07304$ V at $95°C$

However, no glass electrode yet constructed has the theoretical response in all types of test solutions and over the entire practical pH range.

One way of characterizing glass electrode response is *electromotive efficiency*, β_e. This is a fraction less than one, defined by

$$\beta_e = \frac{E_x - E_s}{E^1_x - E^1_s} \qquad \ldots 53$$

E_x and E_s are EMF values of the glass-calomel assembly for solutions X and S respectively and E^1_x and E^1_s are similar EMF values when the glass electrode is replaced by a hydrogen electrode.

If the function β_e is $1 - y$, the error in measured pH is y units per unit difference in pH of solutions X and S. Since individual glass electrodes may have electromotive efficiencies as low as $0 \cdot 995$ an electrode standardized in the standard phthalate buffer at pH $9 \cdot 01$ may read low by more than $0 \cdot 02$ units in the standard borax buffer, pH $(S) = 9 \cdot 18$.

The stem of the glass electrode is usually made of a hard high-resistance glass and the working part of the bulb which is sealed to it is made of special pH responsive glass, e.g., Corning 015 glass.

The pH responsive part must always be completely immersed in solution; if this is not done, the depth of immersion affects EMF values.

The potential of the glass electrode is given by

$$E_g = (E_g)_0 + \frac{RT}{F} \ln a_H \qquad \ldots 54$$

The Corning 015 electrode follows this equation satisfactorily between pH 1 and 9. If the slope of the relation between pH and EMF is found to differ from the theoretical value within this range, experimental technique is at fault. At pH < 1, the Corning 015 glass electrode gives pH values which are too high and above pH 8, values which are too low. In alkaline solutions the error is due to the development of a partial response to cations while in strongly acidic solutions there is evidence of anion penetration into the glass surface.

Glass electrodes made from Lithia glasses have now come into use for alkaline solutions. For some of these the error caused by the presence of sodium ions of concentration 1 mole l^{-1} is negligible for pH values up to 1·2.

The EMF from a glass electrode is not as closely reproducible from day to day as that obtained with a hydrogen electrode; because of this, standardization with a standard buffer solution should be carried out, before and after a series of measurements on unknown solutions.

The following general precautions are usually recommended. The electrode should be kept in distilled water and not allowed to dry out when not in use. It should not be exposed to strongly desiccating solutions or solvents such as absolute alcohol or chromic acid mixture. If by any chance an electrode has undergone such treatment it should be restored by soaking in water. Prolonged contact with alkaline solutions should be avoided and the electrode should be washed in dilute hydrochloric acid after measurements in alkaline solutions.

The causes of pH response of the glass electrode and the theory of electrode behaviour has not been discussed here – a detailed account of these topics is given by Perley.[13] Some common types of commercial glass electrodes are shown in figure 12.

6.3. The quinhydrone electrode

This is the third measuring electrode in order of importance; although not as versatile as the glass electrode it has a much simpler experimental set up. It consists of a bright platinum wire immersed in the test solution previously saturated with quinhydrone. Quinhydrone can be obtained commercially and purified by crystallization from a saturated aqueous solution at 70°C. It is a molecular compound of low solubility made up of a 1 : 1 molecular ratio of

quinone (Q) and hydroquinone (H_2Q)

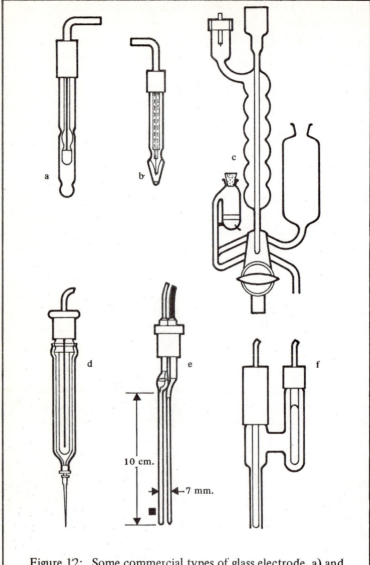

Figure 12: Some commercial types of glass electrode. a) and b), immersion electrodes for general laboratory use; c), MacInnes and Belcher electrode, (widely used in biological experiments) ; d). hypodermic type (Electronic Instruments Ltd.); e), L & N miniature electrodes (glass and calomel); f), Coleman micro glass-calomel cell.

The quinhydrone electrode cell with a calomel reference electrode can be represented by

$$\text{Pt} \left| \begin{array}{c} \text{Solution } X + \text{QH}_2\text{Q} \\ (\text{or } S) \end{array} \right| \begin{array}{c} \text{KCl (aqueous)} \\ 3\cdot5 \text{ N or saturated} \end{array} \left| \text{Hg}_2\text{Cl}_2 \right| \text{Hg} \qquad \dots 55$$

and the cell reaction (neglecting changes at the liquid junction) by

$$\tfrac{1}{2}\text{H}_2\text{Q (in solution } X) + \tfrac{1}{2}\text{Hg}_2\text{Cl}_2 \text{ (insoluble)} + \text{H}_2\text{O (solvent)}$$

$$\rightarrow \text{H}_3\text{O}^+ \text{ (in solution } X) + \text{Cl}^- \text{ (in solution } R)$$
$$+ \text{Hg (insoluble)} + \tfrac{1}{2}\text{Q (in solution } X) \qquad \dots 56$$

For ionic strengths less than $0\cdot1$ M activity coefficient corrections are not necessary in the EMF equation. However, at higher concentration this is no longer so; activity coefficients begin to vary with the nature of the test solution and the electrolytes present in it and pH values differ from those measured with a hydrogen electrode. The difference is called the salt error of a quinhydrone electrode. For practical purposes the salt error is unimportant since it occurs only at high ionic strength where pH ceases to be a well defined physical quantity. However, the quinhydrone electrode has many other limitations; e.g., it cannot be used below a pH of about $8\cdot5$, or in the presence of strong reducing or oxidizing agents since these affect quinone and hydroquinone

Also results are not reliable in presence of proteins owing to some specific chemical interactions and in heterogeneous systems because of selective adsorption of solution.

6.4. The antimony electrode

A stick or pellet of antimony when left on contact with air rapidly aquires a thin layer of antimonous oxide. This oxidation-reduction reaction between antimony and antimonous oxide is the cause of the potential of the antimony electrode. The electrochemical reaction which occurs when an antimony electrode is placed in aqueous solution is

$$\text{Sb}_2\text{O}_3 \text{ (solid)} + 6\text{H}^+ + 6e \rightleftharpoons 2\text{Sb (solid)} + 3\text{H}_2\text{O (liquid)}$$

The antimony electrode is not very accurate. Its potential varies with the condition of the metal, whether cast or electrodeposited, polished or etched. Calibration with a series of standard buffers solutions (carefully chosen for non-complex forming properties) is usually necessary. Even after this the electrode is subject to errors caused by dissolved oxygen, the rate of stirring of the electrolyte and composition of the buffer solution. Various forms of the antimony electrode have been tried, the commonest commercially available one is of cast antimony mounted in a plastic sleeve with only the tip protruding.

Rugged construction and rapid response are the two main advantages of the antimony electrode. It is particularly useful for continuous industrial recording where high accuracy is not required. It is not affected by elevated temperatures and alkaline solutions and it is most convenient for use in humid atmospheres. It is also suitable as an end point indicator in titrations and can be used in cyanide and sulphite solutions where hydrogen and quinhydrone electrodes are inapplicable. It has also been used, under controlled conditions, in the presence of reducing sugars, alkaloids, gelatin and 3% agar solution. However, since water figures in the electrode reaction, the composition of the solvent should be kept constant during a titration. The electrode is also sensitive to oxidizing and reducing agents, to the anions of hydroxy acids (tartares, citrates) which form complexes with antimony and to metaphosphates and oxalates and traces of certain cations (e.g., the cupric ion).

The antimony electrode has various interesting applications in medicine and biology, this is because electrodes of extremely small size function satisfactorily and being immersion electrodes they can be used with cells of the simplest design.

Some properties of the hydrogen, quinhydrone, antimony and glass electrodes have been compared by Bates,[2] the table is given below.

Table 4

Properties of pH responsive electrodes

Property	Hydrogen electrode	Quinhydrone electrode	Antimony electrode	Glass electrode
pH range	unlimited	0–8	0–11	0–13 with corrections
pH response	theoretical	theoretical	variable	nearly theoretical (pH 0–11)
Precision (pH)	±0·001	±0·002	±0·1 (when properly calibrated)	±0·005
Convenience of measurement	low	medium	high	high

Property	Hydrogen electrode	Quinhydrone electrode	Antimony electrode	Glass electrode
Time required for measurement (in minutes)	30–60	5	3	<1
Versatility	low	medium	medium	high
Cost of equipment	medium	low	low	high
Electrical resistance	low	low	low	high
Disadvantages	strong reducing action, air must be excluded	limited pH range, salt error	defective response, not completely reversible	variable asymmetry potential, high resistance, alkaline error
Interferences	oxidizing agents, reducible organic substances noble metal ions, SO_2, CN, unbuffered solutions	proteins, some amines	some oxidizing agents, Cu ions, anions of hydroxy acids	dehydrating solutions, some colloids, fluorides, surface deposits on electrode

6.5. The calomel reference electrode

The ideal reference electrode has a constant and reproducible potential, unaffected by temperature changes and showing little hysteresis. The calomel electrode, the most common reference electrode meets most of these requirements satisfactorily. The basic calomel element can be represented by

$$Cl^- \mid Hg_2Cl_2 \text{ (solution)} \mid Hg \text{ (liquid)}$$

The chloride solution (usually potassium chloride) is saturated with calomel at the surface of the mercury, excess of solid calomel being present to maintain saturation through changes of temperature. If the activity of mercury (I) chloride is thus kept constant, the electrode potential remains constant too.

There are three main types of calomel electrode, these differ in the concentration of potassium chloride used and are called the '0·1 N calomel electrode', 'the 3·5 N calomel electrode' and the 'saturated calomel electrode'. In all three cases it is the salt bridge connecting the reference electrode to the solution of unknown pH which presents most of the problems. The ideal salt bridge would always generate the same diffusion potential, or, no difference of potential, across the liquid junction, but no such perfect bridge is known. A concentrated solution of potassium chloride is commonly used (a saturated solution in the U.S. and usually 3·5 N in European systems) and it functions reasonably well in medium regions of pH but not at high or low pH values.

The properties of the 0·1 N electrode and the saturated calomel electrode are somewhat different. The 0·1 N calomel electrode is reproducible and its potential with respect to the standard hydrogen electrode is almost constant over temperature changes. The EMF of the saturated electrode has a higher temperature coefficient and is subject to larger hysteresis errors. In spite of this, however, the saturated electrode is often preferred to the unsaturated type, because a highly concentrated potassium chloride solution is required in the salt bridge and the use of a saturated electrode avoids the inconvenience of a second liquid junction. However, useful life of saturated electrodes is short when used above 70°C.

There may be a small shift in the potential of calomel electrodes during the first few weeks after its preparation. Bates suggests that this is due to a slow disproportionation

$$Hg_2Cl_2 = Hg + HgCl_2$$

Other causes suggested are change in calomel particle size, slow diffusion processes and physical settling.

Hysteresis is another cause of inaccuracy in calomel electrode measurements. The saturated potassium chloride electrode may show a time lag of several hours in attaining its equilibrium potential after a temperature change. To obtain stabilities of the order of 0·1 mV, 1 to 2 days equilibration time is usually necessary after a temperature change of 20°C.

Polarization can also be a source of error, particularly if the electrode is allowed to take appreciable current. The controlling factor is current density, the maximum anodic current density which a calomel electrode can tolerate over a period of 5 hours has been found to be 15 $\mu A/cm^2$.

A variety of arrangements and cell assemblies have been suggested but there is no preferred design, convenience and mechanical stability being the only considerations. However, a fairly large surface area at the mercury calomel interface is desirable since this makes polarization less likely if current is accidentally

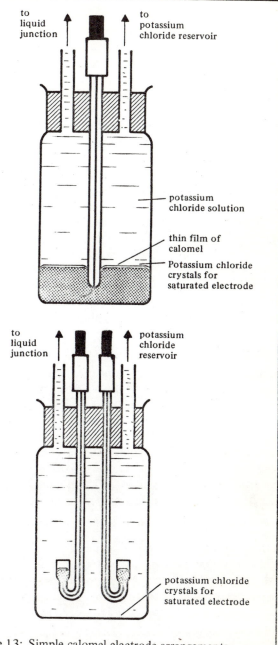

Figure 13: Simple calomel electrode arrangements

taken from the half cell. In addition, the chloride solution should be protected from atmospheric oxygen.

A simple and quite satisfactory arrangement is shown in figure 13.[12] It is important to avoid pockets of potassium chloride solution trapped between the glass walls of the container and the mercury or calomel, these lead to erratic or drifting potentials. Such pockets can be prevented by coating the glass with a silicone fluid, e.g., a 1 per cent solution of Midland Silicones Ltd, MS 200/1000 in carbon tetrachloride or Dow–Corning Silicone Fluid No. 200 also in carbon tetrachloride. The lead wire should be of platinum, and it should not be in contact with the calomel or solution at any time, during construction or later. It can either be sealed into the base of the system so that the mercury rests on it or it can be encased in a glass tube dipping into the mercury.

Figure 14 shows four commercial calomel electrodes the Leeds and Northrup electrode has a replaceable salt-bridge tube so that if necessary different tubes can be used for measurements at different temperatures. The Beckman sleeve type electrode (c) is suitable for emulsions and protein solutions. The Philips electrode uses a piece of porous stone to separate the inner calomel element from the salt-bridge and another to establish the liquid junction.

6.6. The silver–silver chloride electrode

This electrode is highly reproducible and fairly easy to prepare. It was used as a reference by the NBS for the establishment of the pH scale. Its main use, however, is as an inner reference in glass electrodes. When used as a reference electrode in cells without liquid junctions its relative freedom from side effects makes it ideal for the study of the thermodynamic constants of chloride solutions and for the accurate determination of the dissociation constants of weak acids, bases and ampholytes.

The electrochemical reaction occurring between pH 0 and 13 is

$$AgCl \text{ (solid)} + e \rightleftharpoons Ag \text{ (solid)} + Cl^-$$

The silver–silver chloride electrode has three main disadvantages. The first is that its potential is extremely sensitive to bromides in solution; 0·01 mole per cent of bromide ion present as an impurity in potassium chloride solution in which the electrodes are immersed alter the potentials by 0·1 to 0·2 mV. The second important disadvantage is the sensitivity to dissolved oxygen. This is due to the reaction

$$2Ag + 2HCl + \tfrac{1}{2}O_2 = 2AgCl + H_2O$$

69

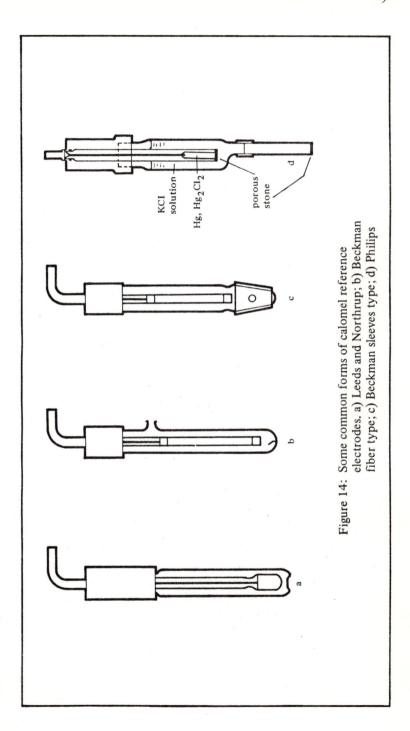

Figure 14: Some common forms of calomel reference electrodes. a) Leeds and Northrup; b) Beckman fiber type; c) Beckman sleeves type; d) Philips

This causes a slight decrease in the concentration of chloride ion in the interstices of the electrode and makes the potential more positive.

The third disadvantage of this electrode is fluctuations in potential caused by an ageing effect during the first 20 to 30 hours after it is prepared. The rate of ageing is dependent on the porosity of the electrode and the stirring of the solution.

Three types of silver–silver chloride electrode, with identical potentials, are in common use. They are 1. the electrolytic type; 2. the thermal-electrolytic type and 3. the thermal type. The first is prepared by electrodeposition of silver on a platinum wire, foil or gauze, and electrolytic conversion of the surface of the carefully washed deposit to silver chloride. In the thermal electrolytic type a paste of silver oxide and water is thermally decomposed to deposit a mass of porous silver on a helix of platinum wire. A part of the silver is then converted to AgCl by electrolysis. The thermal electrode consists of an intimate mixture of silver and silver-chloride formed by the thermal decomposition of a paste of silver oxide, silver chlorate and water.

The thermal electrolytic type of electrode is probably the most widely used. Bates[2] gives details of a method for its preparation.

The standard potentials of the silver-silver chloride electrode, i.e., the EMF in volts of the cell

$$H_2(Pt) \mid H^+Cl^- \mid AgCl \mid Ag$$

have been determined by Bates and Bower.[20]

6.7. Thallium amalgam – thallium (I) chloride electrode

This reference electrode consisting of

$$KCl \, sat \quad TlCl \, (solid) \quad Tl$$

and termed 'thalamid', is for various purposes superior to saturated calomel or silver—silver-chloride electrodes. For example, on heating to 100°C it shows hardly any time lag or hysteresis. 'Thalamid' reference electrodes for use at temperatures as high as 135°C are manufactured by Jenaer Glaswerk Schott und Gen (Mainz). They cannot be used below 0°C as the amalgam solidifies completely below this temperature.

6.8. Ion-specific electrodes

Accurate measurements of the activities of certain cations and anions are essential for an understanding of cellular processes like nerve activity, muscle contraction, enzyme regulation embryonic growth etc. The two most important kinds of electrodes used to measure activities of ions other than hydrogen are:

1. Cation-specific glass electrodes (for potassium and sodium ions).
2. Electrodes incorporating liquid ion-exchanges (for cations and anions).

Cation-specific glass electrodes

Cation sensitive glass electrodes are now very widely used in intracellular studies. Glasses of various compositions have been found to be cation sensitive. An example of a sodium selective glass most common is NAS_{11-18} (containing 11 mole % Na_2O, 18 mole % Al_2O_3 and the rest silica). The four constructions are shown in Fig. 15. In design I the whole micro electrode is made of cation sensitive glass and except for the tip it is painted outside with any glass-adhesive material with good insulating properties. The defect of this design is that the insulating paint often rubs or peels off near the sensitive tip, thus increasing the sensitive area and producing unpredictable changes in electrode potential. Design II has a non sensitive glass micropipette inserted down the tip of a sensitive glass micro electrode like I. The space between pipette and electrode being filled with oil. In this electrode external insulation can still break down but the "functional surface" of sensitive glass is held reasonably constant by the oil and electrode changes are therefore less marked. In design III the sensitive glass microcapillary is placed inside the non sensitive microcapillary and held in place with wax seals. If the lower wax seal remains water tight the electrical performance of this electrode is unaffected. IV is a variation of the same design but has been found more suitable for the construction of micro electrodes with particularly small and sensitive tips. Design V, like designs III and IV, has a sensitive inner and insensitive outer glass capillary but uses a glass — glass seal instead of two wax seals. Its electrical performance is superior to that of the other designs and it also has the most robust construction. Hinke[14] gives detailed methods of construction for all these electrode types. A method for constructing ultra-fine, cation-selective micro-glass electrodes has been described by Kostynk et al.[15]

Liquid ion-exchange electrodes

Liquid ion-exchange electrodes were first used in 1906 but only recently

72

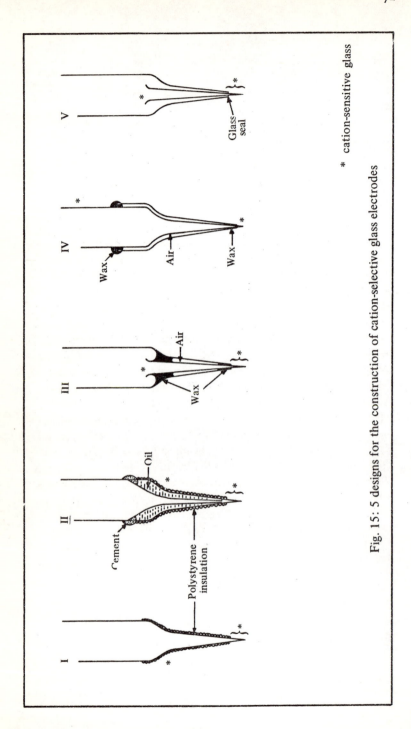

Fig. 15: 5 designs for the construction of cation-selective glass electrodes

* cation-sensitive glass

have they been widely developed.

Early work on the subject reviewed by Beutner[16] shows that
1. Acid material incorporated into an ion-exchanger forms a cation sensitive electrode while basic material forms an anion sensitive electrode.
2. The concentration dependence of the potential follows approximately the Nernst equation (Eqn 10).
3. The membranes responded more to some ions than others i.e. showed ion selectivity.

The electrodes had the following defects
1. Relative transference of anion and cation or perm selectivity was very poor leading to insensitivity at concentrations other than very low ones.
2. Water permeability was high leading to deviations from theorectical behaviour.
3. Ion selectivity between ions of like charge was usually low.

Better liquid ion-exchange materials have since been found which eliminate at least some of these defects. Bonner and Lunney[17] have studied dinonylnapthalene sulphonate (cation exchanger) and kemamine Q-1902-C, National Dairy Prod. Co. Aliquat 336, General Mills Co. (anion exchanger) and obtained under favourable conditions, potentials about 98% of those predicted by Nernst Equation. Sollner and Shean[18] used solutions of lauryl tri alkyl methylamine (Amberlite LA-2) Rohm and Haas Co. and produced membranes with anion specificity excellent permselectivity and low water transfer even at quite high concentrations. Potentials approached Nernst values by a few tenths of millivolts.

Three cation electrodes (for calcium, 'divalent' and for cupric ions) and three anion electrodes (for perchlorate, nitrate and chloride ions) are available commercially, manufactured by Orion Research Inc. Camb. Mass.

The calcium electrode has a limit of detection of 2×10^{3} M, has a tendency to drift and must be shielded carefully from all outside interference. For most accurate results the standard solution must be very similar to the one whose $A_{ca}+$ is being measured. Various authors have studied the properties of these commercial ion exchange electrodes. Their exact formulations are still not available but Orme[19] gives their general composition as follows.

Electrode	Orion Code No.	Type of exchanger
Calcium	92 - 20 - 02	High molecular weight Organo phosphonic acid

con.

Divalent Cation	92 - 32 - 02	High molecular weight Organo phosphonic acid
Chloride	92 - 17 - 02	High molecular weight quaternary ammonium compound in decanol.

REFERENCES

1. Mattock, *pH measurement and titration*, Heywood and Co Ltd. 1961
2. Bates, *Determination of pH; theory and practice*, John Wiley Inc. 1964
3. *Determination of pH of aqueous solutions with the glass electrode*, *ASTM method E 7052*, Philadelphia (1952)
4. Clark and Perley, *Symposium of pH measurement*, ASTM Tech. Pub. 1957
5. Gualandi, et al., *Comm. to the VIIIth. Cong. of Societa Chimica Italiana 1958 sec 16*
6. Severinghaus et al., *J. Appl. Physiol*, 9, 197 (1956)
7. Sigaard et al., *Scan. J. Clin. Lab. Invert.*, 12, 172 (1960)
8. Mattock, *Advances in Analytical Chemistry and Instrumentation*, 2. (Ed. C N Reilley) Wiley—Interscience N.Y. 1963, p. 35
9. Semple, Mattock and Uncles, *J. Biol. Chem.*, 237, 963 (1962)
10. Jacobsen and Mackay, *Nature*, 179, 1239 (1957) and *Lancet* 1924 (1957)
11. Charlton, *Aust. Dental J.* 1, 174 (1956)
12. Dole, *The Glass Electrode*, Chapman and Hall, 1941, pp. 196–202
13. Perley, *Anal. Chem.* 21, 394 (1949)
14. Hinke, *Glass Electrodes for Hydrogen and other cations* (Ed. G Eisenman) Marcel Dekker Inc. N.Y., 1967, p. 464
15. Kostynk, Sorokina and Kholodova, *Glass microelectrodes*, John Wiley Inc. 1969
16. Beutner, *Medical Physics*, 1, Year Book Publ. Chicago 1944
17. Bonner and Lunney, *J. Phys. Chem.*, 70, 1140, 1966
18. Sollner and Shean, *J. Am. Chem. Soc.*, 86, 1901, 1964
19. Orme *Glass Microelectrodes*, John Wiley Inc. 1969
20. Bates and Bower, *J. Res. Natl. Bur. Std. 53*, 293 (1954)
21. Bearden and Thomsen, *Sup. Nuovo. Cim. 5*, 267 (1957)
22. Kolthoff and Strenger, *Volumetric Analysis*, 2nd Edn. Vol. I, Interscience Publishers 1942, pp. 92–93
23. Hitchcock and Taylor, *J. Amer. Chem. Soc. 60*, 2710 (1938)
24. MacInnes, Belcher and Shedlovsky, *J. Amer. Chem. Soc. 60*, 1094 (1938)
25. Bates, Pinching and Smith, *J. Res. Natl. Bur. Std. 45*, 418 (1950)

APPENDIX 1

Values of 2.3026RT/F† and its reciprocal*

Temperature, °C	2·3026RT/F abs. millivolt	F/2·3026RT abs. volt^{-1}
0	54·20	18·45
5	55·19	18·12
10	56·18	17·80
15	57·17	17·49
20	58·17	17·19
25	59·16	16·90
30	60·15	16·63
35	61·14	16·36
40	62·13	16·09
45	63·13	15·84
50	64·12	15·60
55	65·11	15·36
60	66·10	15·13
65	67·09	14·90
70	68·08	14·69
75	69·08	14·48
80	70·07	14·27
85	71·06	14·07
90	72·06	13·88
95	73·04	13·69

* Taken from British Standard 1647:1961
†from ref 21

APPENDIX 2

Values of pH of the primary standard (potassium hydrogen phthalate) at certain temperatures*

Temperature °C	pH
0	4·011
5	4·005
10	4·001
15	4·000
20	4·001
25	4·005
30	4·011
35	4·020
40	4·031
45	4·045
50	4·061
55	4·080
60	4·101
65	4·105
70	4·121
75	4·140
80	4·161
85	4·185
90	4·211
95	4·240

The third decimal figure is not significant, but is included merely to facilitate smooth interpolation.

* Taken from British Standard 1647:1961

APPENDIX 3

Values of pH of NBS standard solutions (other than potassium hydrogen phthalate) at certain temperatures

The data in this table are based on the compilation of NBS standards given in the NBS Letter Circular LC 993 (1950).

$t(°C)$	I	II	III	IV
0	1·67		6·98	9·46
5	1·67		6·95	9·39
10	1·67		6·92	9·33
15	1·67		6·90	9·27
20	1·68		6·88	9·22
25	1·68	3·56	6·86	9·18
30	1·69	3·55	6·85	9·14
35	1·69	3·55	6·84	9·10
40	1·70	3·54	6·84	9·07
45	1·70	3·55	6·83	9·04
50	1·71	3·55	6·83	9·01
55	1·72	3·56	6·84	8·99
60	1·73	3·57	6·84	8·96
70		3·59	6·85	8·92
80		3·61	6·86	8·88
90		3·64	6·86	8·85
95		3·65	6·87	8·83

I: 0·05 M potassium tetroxalate, $KH_3(C_2O_4)_2$. The salt may be purified by crystallization from an aqueous solution below 50°C. It must not be dried above 55°C.

II: Potassium hydrogen tartrate, CO_2K $(CHOH)_2 . CO_2H$ (solution saturated at 25 ± 3°C and decanted from excess salt).

III: 0·025 M potassium dihydrogen phosphate, KH_2PO_4, and 0·025M disodium hydrogen phosphate, Na_2HPO_4. Na_2HPO_4 should be dried for 2 hours at 110–130°C.

IV: 0·01 M borax, $(Na_2B_4O_7 . 10H_2O)$.

APPENDIX 4

List of recommended indicators[22]

Indicator	Transformation range. pH	Colour change
0 – Cresol red (acid range)	0·2–1·8	red–yellow
Thymol blue (acid range)	1·2–2·8	red–yellow
Pentamethoxy red	1·2–3·2	red–violet–colourless
Tropeolin 00	1·3–3·2	red–yellow
2, 4 – Dinitrophenol	2·4–4·0	colourless–yellow
Methyl yellow	2·9–4·0	red–yellow
Methyl orange	3·1–4·4	red–orange
Bromophenol blue	3·0–4·6	yellow–blue violet
Tetrabromophenol blue	3·0–4·6	yellow–blue
Alizarin sodium sulphonate	3·7–5·2	yellow–violet
α – Napthyl red	3·7–5·0	red–yellow
ρ – Ethoxychrysoidine	3·5–5·5	red–yellow
Bromocresol green	4·0–5·6	yellow–blue
Methyl red	4·4–6·2	red–yellow
Bromocresol purple	5·2–6·8	yellow–purple
Chlorophenol red	5·4–6·8	yellow–red
Bromothymol blue	6·2–7·6	yellow–blue
ρ – Nitrophenol	5·0–7·0	colourless–yellow
Azolitmin	5·0–8·0	red–blue
Phenol red	6·4–8·0	yellow–red
Neutral red	6·8–8·0	red–yellow
Rosolic acid	6·8–8·0	yellow–red
Cresol red (alk range)	7·2–8·8	yellow–red
α – Naphtholphthalein	7·3–8·7	rose–green
Tropeolin 000	7·6–8·9	yellow–rose red
Thymol blue (alk range)	8·0–9·6	yellow–blue
Phenolphthalein	8·0–10·0	colourless–red
α – Naphtholbenzein	9·0–11·0	yellow–blue
Thymolphthalein	9·4–10·6	colourless–blue
Nile blue	10·1–11·1	blue–red

Indicator	Transformation range. pH	Colour change
Alizarin yellow	10·0–12·0	yellow–lilac
Salicyl yellow	10·0–12·0	yellow–orange brown
Diazo violet	10·1–12·0	yellow–violet
Tropeolin 0	11·0–13·0	yellow–orange brown
Nitramine	11·0–13·0	colourless–orange brown
Poirrier's blue	11·0–13·0	blue–violet pink
Trinitrobenzoic acid (indicator salt)	12·0–13·4	colourless–orange red

APPENDIX 5a

Characteristics of some commercial electrode glasses at 25°C[2]

Designation of glass or electrode	Composition of glass	Resistance (megohms)	Correction in pH units (to be added)	
			$0.1\,N\,NaOH$	$1N\,NaOH$
Beckman E2	Li_2O, BaO, SiO_2	375	0·03	0·17
Beckman General Purpose	Li_2O, BaO, SiO_2	150	0·43	1·4
Beckman Amber	Li_2O, BaO, SiO_2	550	0·02	0·17
Cambridge Standard	Na_2O, CaO, SiO_2	87	0·7	2·1
Cambridge Alki	Li_2O, BaO, SiO_2	560	0·05	0·25
Corning 015	Na_2O, CaO, SiO_2	90	1·0	2·5
Doran Alkacid	Li_2O, BaO, SiO_2	200	0·07	0·3
Electronic Instruments GHS	Li_2O, Ca_2O, SiO_2	200	0·03	0·16
Ingold U	—	250	0·7	2·3
Ingold T	—	140	0·8	2·2
Ingold UN	Li_2O, SiO_2	30	0·7	1·9
Jena H	—	105	0·8	2·2
Jena U	—	30	0·31	0·7
Jena HT	—	800	0·9	2·2
Jena HA	—	290	0·08	0·25
L & N 'Black Dot'	Li_2O, La_2O_3, SiO_2	70	0·02	0·25
Lengyel 115	Li_2O, BaO, UO_3, SiO_2	15	0·6	1·7
Metrohm H	Li_2O, BaO, SiO_2	1400	0·08	0·15
Metrohm X	$Li_2 O$, CaO, SiO_2	100	0·9	2·2
Metrohm U	Li_2O, BaO, SiO_2	500	0·08	0·25

APPENDIX 5b

pH values of aqueous solutions recommended for calibration of glass electrodes[+] [23] [24] [25]

Composition in moles/litre of solution	pH values at various temperatures			Reference
	12°C	25°C	38°C	
$0.1\ KH_3(C_2O_4)_2 . 2H_2O$	—	1·48	1·50	1
$0.01\ HCl + 0.09\ KCl$	—	2·07	2·08	1
$0.05\ HOOC . C_6H_4 . COOK$ (primary standard)	4·000	4·005	4·026	By definition
$0.1\ CH_3 . COOH + 0.1\ CH_3COONa$*	4·65	4·64	4·65	1, 2
$0.01\ CH_3COOH + 0.01\ CH_3COONa$*	4·71	4·70	4·72	2
$0.025\ KH_2PO_4 + 0.025\ Na_2HPO_4$	—	6·85	6·84	1
$0.05\ Na_2B_4O_7 . 10H_2O$	—	9·18	9·07	1
$0.025\ NaHCO_3 + 0.025\ Na_2CO_3$	—	10·00	—	3

* Prepared from pure acetic acid, diluted and half neutralized with sodium hydroxide: it should not be prepared from sodium acetate.
[+] Taken from British Standard 1674:1961

INDEX

COMMERCIALLY AVAILABLE EQUIPMENT

A. Tables of pH METERS manufactured by:

1.	Analytical Measurements	(U.S.)
2.	Beckman	(U.S.)
3.	Biolyon	(French)
4.	Cambridge	(British)
5.	Cenco	(Dutch)
6.	Chandos Products	(British)
7.	Coleman	(U.S.)
8.	Corning	(U.S.)
9.	Derritron	(British)
10.	Electronic Instruments	
11.	Horiba	(Japanese)
12.	Lapine	(U.S.)
13.	Leeds & Northrup	(U.S.)
14.	Ludwig Seebold	(Austrian)
15.	Meci & Minipoint	(French)
16.	Metrohm	(Swiss)
17.	Ogawa Seiki Co.	(Japanese)
18.	Phillips	(Dutch)
19.	Photovolt Corp	(U.S.)
20.	W.G. Pye	(British)
21.	Radiometer	(Danish)
22.	Sargent	(U.S.)
23.	WTW	(W. German)

B. Table of AUTOMATIC TITRATORS manufactured by:

1. Beckman
2. Coleman
3. Doran
4. EIL
5. Pye
6. Radiometer

1   ANALYTICAL MEASUREMENTS

Type Cat.No.	Range	Power	Temp. Comp.	Recorder Output	Special Features	Reproducibility
Redox pH meter	0-14 pH ±1400 mV	line operated	0-100°C	None	Equipped with protected pH probe unit	.02pH
Pocket pH meter	0 to 14 pH 2 to 12 pH	battery				
Big Scale pH meter	0 to 14 pH	line operated			"	
Recording pH meter 3 RS		line operated	0-100°C	Automatic strip recorder connected to pH meter speed: 1inch per hour	"	
pH recorder controller		line operated		Stripchart recorder connected to pH meter	Equipped with upper and lower adjustable contacts for "on/off" control, for operating external equipment when pH value exceeds preset limits	

2   BECKMAN INSTRUMENTS LTD

Type Cat. No.	Range pH	±mV	Accuracy	Power	Scale Expander	Temp. Comp.
Model 72 S.72001	0–14 pH	± 500 or ±1000 by displacing zero	±0.3pH or ±0.1pH by standardising with buffer in region or sample pH	230V(±10%) 50/60Hz6w	none	manual 0–100°C
Zeromatic pH meter 151400	0–14 pH	0–1400mV or –700 to +700 mV	±0.05pH or 5 mV	110V, 125V, 225V or 250V, ±10% 50–60Hz	none	manual 0–100°C or automatic using thermocompen sator No.39096
Expandomatic pH meter 151500	normal 0–14 pH expanded 0–2 pH	normal 0–1400mV expanded 0–200 mV	standard ±0.05pH or 5mV expanded ±0.01pH or 1mV	115V or 230V ±10% 50Hz	standard scale discrimination of ±0.025pH is expanded to ±0.0025pH	0–100°C manual of automatic using thermocompensator No.39096
Research pH meter S101901	–0.5 to 14.5	–50 to 1450mV or +50 to –1450 mV	±0.001 pH over any 3pH or ±0.0037pH over whole range	105V to 125V 50/60 Hz	none	manual 0–100°C adjustable to nearest 0.05°C of sample temp.

BECKMAN INSTRUMENTS (Continued)

Reprodu-cibility	Recorder Output	Special Features	Applications
±0.05 pH		With the combination electrode which is supplied as a standard, pH measurements can be done below 50°C. Above this a large selection of Beckman glass metallic and reference electrodes is available.	Suitable for rapid and accurate pH measurements, useful for pH titrations, Karl Fischer titrations and in food and beverage processing, in pharmaceutical laboratories and in petroleum, pulp and paper processing.
	Adjustable with appropriate resistances to suit various potentiometric and current operated recorders	Large pivotless taut-band meter with scale of 21 cms. operates well on low currents. Mirrored scale. There is automatic zero standardizations once a second. So there is no need for manual correction of amplifier drift.	A compact mains operated instrument for making precise pH and mV determination
	Adjustable with appropriate resistances to suit various potentiometric and current operated recorders		A versatile laboratory pH meter. Suitable for rapid measurements requiring a high degree of accuracy and discrimination.
	Any potentiometer operated recorder with floating input and spans from 1-100 mV may be used by means of a recorder adaptor plug assembly	Linear (3.05 meters) slide wire provides null balance and very high resolution in measurements. Readout dial allows rapid finger-tip adjustment coupled directly to slide wire. Readout is graduated in steps of 0.002pH with interpolation of 0.0005 pH units.	High precision work

3 BIOLYN

Type Cat.No.	Range pH	±mV	Accuracy	Power	Scale Expander	Temp. Comp.	Reproducibility
Cosmos	0–14 or ±1.5 pH with displaceable zero from 0–14	±400 mV or ±800 mV or ±2000mV with incorp. displacer	normal ±0.02pH expanded ±0.005pH	127–220V 40 to 60Hz insensitive to variation of ±20%	see range	manual 0–100°C	±0.0005pH
Horus	0–14 or ±1.5pH with displacable zero from 0–14	±400mV or ± 800 mV or ±2000 mV with incorp. displacer	normal ±0.02pH expanded ±0.005pH	127–220V 40 to 60Hz insensitive to variation of ±20%	see range	As above	±0.0005pH or ±0.25 mV
T 43			±0.03 pH or ±2 mV	115 and 230V 40 to 60Hz ±0.05 pH or ±3 mV for variation of ±10% in input tension		As above	0.005pH
Pim (new portable)	0–7 or 7–14		reading ±0.02pH measuring ±0.03pH	2 small batteries of 49V 1 small battery of 1.5 V	None	As above	±0.01 pH
S 53 (new)	normal 0 to 14 expanded 5.5 to 8.5		normal ±0.03pH or 1.8mV expanded ±0.01 pH or 0.6mV	110 or 220V 40 to 60 Hz 0.01pH for a variation of 10% in input tension	See range	As above	±0.005pH or ±1 mV

BIOLYN (continued)

Recorder Ouput	Special Features	Application
not listed	Electrical circuit entirely transistorized. Excellent temp. compensation. Anti-parallax mirror. Can be used in conjunction with a variety of electrodes including metallic ones	Suitable for high precision work where good temperature control is required.
not listed	Electrical circuit entirely transistorized. Excellent temp. compensation. Anti-parallax mirror. Can be used in conjunction with a variety of electrodes including metallic ones	Suitable for high precision work where good temperature control is required.
not listed	Direct reading, lightweight pH meter with shockproof watertight galvanometer	High precision instrument suitable for use under conditions of high humidity.
not listed	Anti-parallax mirror. Galvanometer scale of 85 mm. Portable, entirely transistorized, can be used in conjunction with all electrodes normally used with pH meters	Suitable for use under conditions of extreme fluctuations in temperature
not listed	Direct continuous reading, two different inputs for pH and mV. Lightweight pH meter. Highly shock resistant	High precision instrument suitable for use under conditions of high humidity

4 CAMBRIDGE INDUSTRIAL INSTRUMENTS LIMITED

Type Cat. No.	Range	Accuracy	Power	Temp.Comp.	Recorder output	Applications
Cambridge 44239	0–14 pH –400 to + 1000mV	±0.05pH units	AC mains 200–250 V, 50 C/S	10 –100°C	Available	Laboratory or industrial pH meter

5 CENCO

Type Cat. No.	Range pH ±mV	Accuracy	Power	Temp. Comp.	Special Features	Applications
34462 CENCO Recording	2–12 with 0.02 divisions	0.1 pH	220 Volts, 50 cycles, 25 watts.	0°C–100°C manual	Simple to operate suitable for pH monitoring	Suitable for laboratories, plants and classrooms
34461, Bigscale pH meter	0.14 with 0.1 divisions	±0.005pH	220 volts 50/60 cycles 10 watt	0°C–100°C	One knob control Readability 0.02pH	spear type probe enables the pH of semi-solids and soil to be measured
34463	0–14 ±1400	0.05pH 3 mV	220 volts 50 cycles 25 watts	0°–100°C manual	Simultaneous Millivolt and pH readings without changing electrodes	Suitable for lectures and demonstrations
34460 Pocket pH meter	0–13.5	0.1 pH	Battery	for 2–12 pH up to 100°C	checkpoint reference system	

6 CHANDOS PRODUCTS (SCIENTIFIC) LTD

Cat No Type	Range pH	Accuracy	Power	Temp. Comp.	Recorder Output	Special Features and Applications
Miniature portable pH meter A 53	0 – 14	0.2 pH	battery	0 – 100°C	None	The electronic buffer position on the main switch allows long periods of use between successive standardizations with a buffer solution
Battery operated dual scale meter A54	0 – 8 or 6 – 14	0.1 pH	battery	0–100°C	None	As above
dual ranged pH test meter A 55	0 – 14 or ± 3 pH from centre	0.1 pH or 0.05 pH for second range	battery	0 – 100°C	None	As above
Standard pH meter A51	0 – 14 or ± 3 from centre	0.2 pH for 1st range 0.1 pH for 2nd range	battery	0 – 100°C	None	A 'check' position is included in the main rotary switch and this can be used (if the 'check' point of the electrode is known) to make measurements at any temperature up to 100°C without using a buffer solution
Mains operated standard pH meter A 52	As above	As above	Line operated	0 – 100°C	can be used with a pen recorder	As above

7 COLEMAN

Type Cat. No.	Range pH	mV	Accuracy	Power	Scale Expander
Metrion III 28–030 Coleman Model 28B	0–14		±0.05pH	95–125 volts 50 or 60 cycles AC 10 watts Also available for 220 volts 50 cycles A.C.	none
Metrion IV 28–040 Coleman Model 28B	0–14	–1400 to +1400	±0.05 pH accurate to 2% full scale	95–125 volts 50 or 60 cycles A.C. 10 watts, stabilized against voltage variations from 95–125 volts	none
Model 37 A 37–002	0–14	0 to ±1500	±0.001 pH at standardization point. ±0.005pH over any 3 pH span without additional standardization ± 1.0 mV	100–130 volts 50/60 cycles A.C. or one nickel cadmium rechargeable battery, 3 watts	
Coleman Model 38 38–000		±1400	normal mode ±0.03pH on mV range. ±0.5%F.S. Expanded mode ±0.005 pH	100–130 volts 50/60 cycles A.C. 15 watts	0–14pH in 1.0 pH steps
Coleman Model 39 39–000	0–14	–1400 +1400	±0.03pH 0.5% full scale	100–130 volts 50/60 cycles A.C. 12 watts	

COLEMAN (Continued)

Temp. Comp.	Reproducibility	Special features and Application
0–100°C	0.02pH	Simple routine pH measurement in industrial, clinical and academic applications
0–100°C		Mirrored scale. Automatic titrator circuit.
	±0.001	Portable with field effect transistor input. Null–balance circuitry. Can be operated in vertical, horizontal or tilted position.
manual or automatic	Normal 0.01pH Expanded 0.001 pH	Taut suspension friction free meter movement provides reading free from friction error, greater resistance to shock
manual or automatic	±0.01pH	

8   CORNING

Type Cat No	Range pH	mv	Accuracy	Power	Scale Expand
Corning Model 5 475000	0–14	± 700	± .1 pH	line operated 105–125v A.C. 50/60 C/s	none
Corning Model 7 475007	0–14	0 ± 1400	± 0.5 if pH of sample is within 4 pH units of calibration point	line	none
Model 10 expand scale pH meter 475010	normal 0–14 expanded to any 3 pH	normal 0 to ± 1400 expanded to 0 to ± 300	normal ± 14 mv expanded ± 3 mv	105–125v A.C. 50–60 Hz 6 watts	normal 0.05 pH expand 0.01 pH
Model 10C pH controller 475080	normal 0–14 expanded any 3 pH	normal 0 to ±1400 expanded 0 to ± 300	normal ± 0.05 pH within 4 pH units of calibration point expanded ± 0.01 pH	line 105–125v A.C. 50/60 Hz 6watts	gives smallest division equal to 0.01 pH
Model 12 research pH meter 475012	0–14 expanded 1 pH for full scale deflection	± 1400	normal ± 0.05 pH expanded (1) at buffer point ± 0.002 pH (2) with buffer and sample in same range ± 0.005 (3) under conditions other than the above ± 0.01 pH	line stabilized for variations from 105–135vAC. easily modified fur use with 220v	1 pH unit gives full scale deflection smallest division is 0.005 pH

CORNING (Continued)

Temp Compensation	Reproducibility	Recorder Output	Special Features	Applications
automatic 0–100°C	± 0.05 pH	10 mv for full scale deflection		suitable for educational or industrial laboratory
automatic and manual 0–100°C	± 0.02 pH	As above		
automatic 0–100°C for normal range 0–100°C for expanded range 5.5pH–8.5pH manual 0–100°C for all ranges	± 0.005 pH for expanded range		simple to operate	an accurate expanded scale pH meter and a precise pH controller
0–100°C automatic and manual	normal ± 0.02 expanded ± 0.005	variable 10–100 mv for full scale deflection		
manual 0–50°C optional automatic temp. compensator probe works on normal and expanded ranges	± 0.002 pH or 0.2 mv on expanded scale	variable 10–100 mv easily modified for galvanometer type of recorder Karl Fischer output: 5μA		for oxidation reduction measurements and for Karl Fischer and other polarized electrode titrations

9

DERRITRON INSTRUMENTS LTD
(formerly Doran Instrument Co Ltd)

Type Cat No	pH indicator M 4940	Mini pH meter M 4971	Universal pH meter M 4982	Universal pH meter and DC potentiometer M 4989
Range pH mv	0–14 0–1000	not specified	0–14 0 to ± 1400	
Accuracy	05 pH	'sufficient for general lab use'	0.01 pH	accuracy meets BS specifications for 'pH scale' 1647 (1961)
Power	A.C. line	self- contained batteries	2V accumulator and 2 x 9V batteries	2V accumulator and 2 x 9V batteries
Temp Comp	manual		automatic	automatic

Recorder Output	none	none	not specified	
Special Features	very stable circuit, simple to use	Simple and robust	low grid current	
Application	routine and titration work	factory and outdoor use	general purpose laboratory instrument	consists of an electro- meter valve potentiometer calibrated in pH units and millivolts

10 ELECTRONIC INSTRUMENTS LIMITED

Cat No Type	Range pH	Accuracy	Power	Temp Compensation
Portable pH meter Model 23A	(1) 0–14 (2) ΔpH scale covers a deviation range of ± 3.5 pH about an arbitrary normal (3) 0–800 mV	.1 pH on 0–14 pH scale 0.02 on ΔpH scale 10 mv on 0–800 mV scale	line 100–130 V 200–250 V A.C. at either 50 or 60 C/s	automatic 0–100°C
Vibret Laboratory pH meter Model 3920	0–14 0–1.4 with backoff in steps of 1 pH also ±0–1400 mV and ±0–140 mV	± 0.05 pH on 0– pH scale, ± 0.005 pH on 0–1.4 pH ± 0.5 mV on 0–140 mV	100–120 V or 200–250 V 50 Hz 25watts (60 Hz model available)	automatic or manual over –5°C–100°C
Vibron pH meter 39A	0–14 and 14 expanded ranges of 1.4 pH or 0–1400 mv of any polarity	0.005 pH on expanded scale		manual 0–100°C automatic 0–100°C
Portable Model 30C	0–8 or 6–14	± 0.1 pH or ± 10 mV	9V dry battery	automatic 0–100°C using probe 20
Laboratory type Model 38A	0–8 or 0–14 or 0–800 mV and 600–1400 mV (negative potentials only)	± .1 pH or ± 10 mV	100 to 250 V A.C. 50/60 C/s or 18 V DC from batteries	manual and provision for automatic over 0–100°C
Vibret laboratory pH meter Model 46A	0–14 and 0–2.8 with back off in 6 steps of 2pH	± 0.05 pH in 0–14 range or ± 0.01 pH on 2.8 pH ranges	either 200– 250V or 100–120VA.C. 50 C/s	automatic 0–100°C
Vibron blood pH meter Model 48B	6.6–8.0 and 3 – 10	discrimination of 0.005 pH for first range	not specified	not specified

ELECTRONIC INSTRUMENTS LIMITED (Continued)

Recorder Output	Special Features	Application
terminals to recorders of sensitivity up to 100 μA per pH unit or up to 20 mV/pH unit and overall resistance up to 200 ohms	meter conforms to latest BS limits for precision meters Has very stable DC amplifier. Zero drift under normal conditions not exceeding ± 0.02 over 24 hours	For direct reading and for recording in the laboratory and industrially. Internal check circuit for testing adjustment without disturbing electrodes enables calibration to be kept within 0.05 pH in normal industrial use. Also suitable as direct reading milli volt meter in redox measurements.
1.4 mA into upto 5000 ohms	Kare Fischer 5 μA output provided. Stability typically 0.005 pH per day. Shock proof	Advanced general purpose laboratory pH meter
0-700 μA into 5k ohms	Uses special vibrating capacitor which gives excellent zero stability. (0.005 pH in 12 hours)	A versatile instrument suitable, on expanded range, for highly accurate work.
not specified	stability = 0.05 pH/day or 3 mV/day after one hour warm up period	routine measurements
not specified	Instrument is suitable for measuring negative electrode potentials between 0-1400 mV Has moveable electrode clamp.	convenient for titrations
1.4 mA full scale, maximum permissible recorder impedence 5KΩ	Stable and robust circuit, can be used for measuring positive and negative potentials	A precision laboratory instrument
700 μA full scale into 5KΩ	zero stability better than 0.005 pH in 12 hours	A precision indicator for pO_2, pCo_2 as well as pH

11 HORIBA

Type	Range pH	mv	Accuracy	Power	Scale Expander	Temp.Comp.
Horiba F-5	0 to 14	± 1400	Normal Scale ±0.03 pH or ± 5 mV expanded scale ±0.01pH or ± 1 mV	A.C. 220 volts (±10%) 50-60 cycles, approx. 7 watts	Any 2pH span within the range of pH 1 to 15 Any 200mV span within the range of +800 to -800 mV	0 to 100°C either manually -by a dial -or auto- matically by an attached thermo- compen- sator
M-5	0-14	0 to ±1400 0 to ±700 1000 to 1400	±0.03 pH or ±5 mV	A.C. 220 V (± 10%) 50-60 cycles (common) approx. 7 watts		0 to 100°C
Compact D-5	0 to 14		±0.05pH	(1) Battery: 12 UM-3A (1.5V unit) dry batteries are built-in (2) Line-operation: A.C. 220 V (± 10%) 50-60 cycles 1 watt		0 to 100°C
Horiba H-5	0 to 14		±0.05pH	A.C. 220 volts ±10%, 50-60 cycles approx 1 watt		0 to 100°C is made automatically

HORIBA (Continued)

Recorder Output	Special Feature	Applications
Two connections, for a potentio-metric recor-der with a span of 10mV and a current recorder with a span of 2.8 mA	A variety of pH electrodes can be used with this instrument	Can be used as a convenient general purpose pH meter.
Span of 10mV or less, or for a current re-corder with a span of 2.8mA or less. (in each case the output is zero at pH 7)	Self-balancing amplifier maintains a balance between input and output. A combination electrode is employed. The temperature range for the electrode is $0 \sim 50^{\circ}$C	
	D.C./A.C. two-way system, the amplifier is fully transistorized and can be operated for the period of 3 months in normal use. Enables pH measurement out of doors	
	A D.C. amplifier provides economical pH measurements with line operation	Operated easily in the classroom

12 LA PINE SCIENTIFIC CO

Type Cat.No.	Range	Accuracy	Power	Reproducibility	Applications
Lapine Portable Battery operated pH meter 203.92	0–14 pH 0 to ± 700 mV	±0.1 pH	Battery one flashlight D–cell and one 2U6 transistor radiocell	0.05 pH	Designed for routine laboratory applications: soil testing, food, beverage processing and for use in pharmaceutical, commercial or hospital laboratories

13   LEEDS & NORTHRUP

Type Cat. No.	Range pH	Range ±mV	Accuracy	Scale Expander	Temp. Comp.	Reproducibility	Special Features	Application
7401-A$_2$	0–14	0 to ±700 0 to ±1400	±0.05pH			0.02pH	Readability, stability and speed of response	General laboratory purpose pH meter
7405-A$_2$	0–14	0 to ±1400	±0.01pH	6 to 8pH graduations may cover any 2-pH span		0.003pH	A 10- μA polarising current is provided for dead-stop end-point measurement. Fast response	Suitable for quality control or in biomedical pharmaceutical or chemical research
7407-A$_2$	0–14	0 to 1400	0.005pH	200 mV expanded range		0.002pH	Suitable for pH measurement of blood samples and other body or biological fluids.	General research. Precise clinical and research measurement
7403-A$_2$	0–14	0 to ±700 and 0 to ±1400	±0.05pH	2-pH span and 200 -mV span	Correct on all 2-pH spans; independent of the iso-potential value of the electrode	0.001pH	Direct change from 'expanded' to 'full-range', scales without recalibrating.	For high accuracy requirements in research, medicine, and industry
Speedomax Indicators Recorders and Controllers			± 0.3% of span for recorder (0.03 pH for 2-12 pH range)		Automatic		Indicating recorder (round or strip chart recorder). Built in high impedance amplifier receives signals direct from pH electrodes	

14 LUDWIG SIEBOLD

Type Cat No	Glass electrode pH meter Type GBA
Range (pH)	1-13
Accuracy	± .1 pH
Power	2 batteries of 9 and 1.5 volts
Temp Compensation	automatic 10°C to 80°C
Recorder Output	not specified
Special Features	A compact pH meter. It is equipped with a shock-resistant unit-probe and has a robust, acid proof plastic case. Electrode probe can be wiped if necessary.
Applications	Suitable for measuring the pH of soil samples, food stuffs like butter, yogourt, chocolate etc., and of cosmetic creams and pastes and wood pulp. Its shock-resistant properties make it also suitable for use in pickling baths, plating solutions, tanning, dyeing and bleaching, control of boiler feed water and in fermentation processes. It is also useful for general process and quality control.

15 MECI AND MINIPOINT

Type Cat No	Range pH / mv	Accuracy	Power	Scale Expander
Indicator VS2	3 inter- changeable scales 0–6 0–600 4–10 400–1000 8–14 800–1400	0.01 pH	line, 115V, 50 Hz	None
Recorder balistic	not specified	not specified	45 volts HT battery. 6.3v transformer for 110V, 50 Hz	none

Note: MECI also manufacture Speedomax and Minipoint under licence from Leeds and Northrup (q.v.)

Temp Comp	Reproducibility	Recorder Output	Special Features	Applications
0–80°C	3%	none		Designed for research and industrial laboratories
available but range not specified	not specified	trace on 250mm wide graph. Speed 50mm/hr modified on request	automatic calibration every 45 min	Designed for industrial use. Gives continuous record of pH. Intended for attachment to some manufacturers control systems.

16 METROHM

Type Cat.No.	Range pH	mV	Accuracy	Power	Scale Expander	Temp.Comp.	Reproducibility
E 280 A	0–14	∓500, 0 to ±1000	0.1 pH	Battery 2x1.35V, 2x15 V		0–100°C	
E 444	0–14	∓500	0.05pH within 4pH of the calibration –point	accumulator (with built –in charger)		0–100°C graduation 10°C	
E 350 B	0–14	∓700 with zero–point shift to get 0 to –1400 mV or 0 to +1400 mV	0.05pH (within 4pH units from calibration value) 10mV absolute	mains–operated 220/250 V 50 C/s A.C.		0–100°C	
E 396 B	0– 8 6–14	–150 to +1000 and +150 to –1000	0.02pH, 5mV absolute	110–125V 220 to 240V ±10% frequency 40–60 C/s.		can be set to an accuracy of 2°C over a range of 0–100°C.	0.01pH
E 300 B pH/mV meter	0–14 5.6–8.4	∓700, ∓1400		220/250V, 50 C/s A.C.	The 2.8–pH fine range can be shifted to any position between 0–14 pH		0.005pH (fine range

METROHM (Continued)

Recorder Output	Special Feature	Application
	Battery capacity approx. 100 working hours, portable	For open air work and for carrying out check measurements in a factory.
	Portable pH meter fixed-point control with built-in battery charger. Battery-life: approx. 36 hours	Both E 280 A and E 444 use a DC amplifier with very high input impedance for measurements with glass electrodes.
	Built-in polarisation current source for $3\mu A$, voltametric titration can be carried out without the need for additional apparatus.	Useful laboratory instrument
For recorders with sensitivity adjustable between 20 and 50 mV	very stable	
	Zero-point stable AC amplifier provides exceptional stability over long periods.	wide fine range for precision work.

17 OGAWA SEIKI CO LTD

Type Cat. No.	Range	Accuracy	Power	Scale Expander	Temp. Comp.	Reproducibility	Recorder Output	Special Features	Applications
Glass electrode pH meter Toa Dempa	0-14pH 0 to ±1400 mV	±0.03pH 5mV	line operated		automatic 0-100°C	not specified	has output terminal for recorder	High stability, direct reading meter	Acid base titrations for measurement of acidity of soil and soil solutions and for pH of blood, wine etc.
Portable pH meter Toa Dempa	1-13pH	±0.1pH	battery		manual 0-100°C	not specified		Light weight	Suitable for use in the field or pH of industrial water and drinking water.
TOA Model Omnimeter									The unit does the work which a medium class single feature unit would do in each method of measurement.

consists of Photo electric colourimeter, Polarograph, pH meter, conductivity meter, electrolysis, coulombmeter

pH meter has following characteristics

Range	Accuracy	Power		Temp. Comp.	Reproducibility	Recorder Output
0-14pH	±0.05pH	subject to order		manual 0-65°C	not specified	available

18 PHILIPS

Type Cat. No.	Range pH	Accuracy	Power	Temp. Comp.	Reproducibility	Recorder Output	Applications
Battery Portable PR9401	0 to 14	$\pm$.02pH	Battery	Manual 0–100°C	$\pm$0.02 pH	not specified	routine measurements and sample analysis
Industrial PR9402/01		$\pm$.02pH		automatic	$\pm$0.02 pH	not specified	Process control: chemical, pharmaceutical, leather, oil, paper, textile effluent
Direct reading PR9403/01	25 different ranges, spans as low as 2pH, from 2 to 14 pH	min .01pH max .06pH		automatic if Nickel resistance thermometer connected	min .01pH max .06pH	not specified	Blood pH measurement etc.
Mains Portable		0,1 pH	line	Manual 0–100°C	0.1 pH	not specified	routine measurements and sample analysis

19 PHOTOVOLT CORPORATION

Type Cat. No.	Range	Power	Scale Expander	Temp. Comp.	Reprodu- cibility	Recorder Output	Special Features
High Precision 111	0–14 pH ±700 mV and ±1400mV	line operated	none	manual 0–100°C or by ex- ternal automatic probe	0.02 pH	Adjustable Recorder Outlet	Solid state design, illuminated push button controls
OmniRange Model 180	0–14 pH 7 expan- ded ranges each of 2 pH units	battery	7 expan- ded ranges each of 2pH units		0.01 pH		

20 W.G.PYE

Type Cat.No.	pH	Range ±mV	Accuracy	Power	Scale Expander	Temp. Comp.	Reproducibility	Recorder Output	Special Feature	Applications
Portable 11071	2-12 or 0-14 by buffer control		± 1%	Batteries Ever-Ready LP/U2(3) B115(3)	none	none	not specified	not specified	Easily portable	For periodic checking of process solutions in factories or in the field
Model 290 11290	0-14 and 14 ranges of 1.4 pH for f.s.d.	±1400 and 14 ranges of 140mV for f.s.d.	0.01pH	100 to 110V or 200 to 250V 50 to 60 Hz	span of 1.4pH gives smallest division of 0.01pH	manual 0-100°C automatic -5 to 100°C using resistance thermometer (Cat. No. 625)	not specified	not specified	Discrimination of 0.002pH. Buffer control with memory dial. High stability, very high input impedance.	Accurate laboratory pH meter.
Model 291	0-14 or 0-2.8 covering any part of the 0-14.8pH span			100 to 120V or 200 to 250V 50 to 60 Hz	7 in. scale covers 0-14 pH and in expanded form 0-2.8 pH	manual 0-100°C automatic -5 to 100°C using resistance thermometer (Cat. No. 625)	not specified	not specified	Very high input impedance	
Model 78 11078	0-14 3.5-10.5	±175mV increased to ±1175mV in steps of 200mV		110 to 120V and 200 to 250V 50 to 60 Hz	none	manual 0-100°C automatic 0-100°C for E07 electrode, using resistance thermometer Cat. No. 625	not specified	not specified	Supplied complete with combined glass and reference electrode 401 E7	
Model 79 11079	0-14pH	±350 mV		line 110-120V and 200-250V or using A.C./D.C. converter (Cat. No. 11280) can be operated by batteries		manual 0-100°C automatic 0-100°C using Res. thermometer Cat. No. 623			Electrical zero of 7. Good stability. Supplied with combined glass and reference electrode 401 E7	

21 RADIOMETER

Type Cat No	Range pH	mv	Accuracy	Power	Scale Expander	Temp. Comp.
pHM4	0–14	–1500 to +1500	$\pm$ .002 pH $\pm$ .2 mv $\pm$ .5%	battery	none	0 – 40°C manual
pHM22	0–8 and 6–14		$\pm$ .02 pH	line operated	pHA using 621: 6.6 – 8.0 pH pHA using 630P: 1.4 pH total scale	
pHM24	0–8 and 6–14		.05 to .02 pH	built in miniature batteries	none	0 – 100°C manual
pHM26	0–14 expanded to 1.4 pH units for full scale deflection	0 to $\pm$1400 expanded 140 mv	.5 pH or .005 pH	line operated, insensitive to variations of $\pm$ 15%	1.4 pH units give full scale deflection	manual 0 – 100°C automatic (using temp. compensator T501) –20°C to +120°C
pHM27	0–12 pH expanded 6.8 to 8.2	0 to $\pm$1200 mv expanded 0 to $\pm$120mv		line operated	built in for blood measurement	manual 0 – 100°C automatic 0 – 120°C
pHM28	0–10 6–14	–800 to +200 – 400 to + 1600	$\pm$ .02 pH	line 115V or 220V	none	manual 0–100°
pHM29	0–10 pH 4–14 pH		.05 to .02 pH	battery or line		
pHM31	2–10 pH		.05 to .02 pH	line operated		
pHM32	2–10 pH		.05	line operated		

RADIOMETER (Continued)

Reproducibility	Recorder Output	Special Features	Applications
± .1 mv ± .001 pH	none	Electrical zero at 6.4 pH, mirror backed scale	General and precision pH meter, pH measurements in blood
		Electrical zero at 7 pH, mirror backed scale	Precision chopper stabilized laboratory pH meter. Suitable for use under humid conditions.
	none	Adjustable electrical zero, mirror backed scale	Portable pH meter
± .01 pH or on expanded scale ± .002	100 μA/pH	A special meter scale and polarization voltage of 280 mv is incorporated for accurate deadstop endpoint titrations. Mirror backed taut band suspended meter.	High precision laboratory instrument. Radiometer titration unit type TTT11. converts it into a highly accurate automatic titrator. Suitable for use under humid conditions
± 0.01 pH or ± 0.002 pH	Current .1mA/pH impedence IM ohm voltage 10 mV/pH impedence 100 ohms	Mirror backed scale Useful under conditions of high humidity	Chopper stabilized laboratory instrument with meters of pH, mV, Pco2 and standard bicarbonate convertible to automatic titrator using TTT11
± .02 pH	1 mA/pH 700 ohms	Electrical zero at 8 pH. Buffer adjustment control possible.	Laboratory pH meter, convertible to automatic titrator using TTT11
			Lightweight laboratory pH meter
			Industrial direct reading meter with watertight cabinet, calibration in mV available. Suitable for humid conditions
			Industrial pH meter, no indicating meter, watertight cabinet. Suitable for use under humid conditions

22 SARGENT

Cat No Type	Models PB and PL portable pH meters		

Range	pH mv
	
	0–14 –700 to
	          ·700 or
	          0 to
	          ± 1400
		Temp Comp	manual –2°C
			automatic
			0° to 60°
			($\pm$ 2°C)
			60° to 80°
			($\pm$ 5°C)

Accuracy	Absolute accuracy at meter is 0.05 pH or $\pm$ 5 mV, at recorder outlet it is $\pm$.025 pH or $\pm$ 2.5 mV	Recorder Output	Recommended for use with recorders of 5 mv to 250 mv full scale range
Power	Model PL operates from mains using Zener referenced supply Model PB has Mercury battery supply, 8 volts providing 800 hours use	Special Features	A 'buffer adjust' control allows accurate adjustment to fit electrode response. After initial warm up period. Electronic circuit drift is 0.02 pH per hour. It is a rugged instrument with shock - proof case and moisture proof transistorized circuit and taut band suspension meter. Jacks for connecting glass electrodes accept Sargent, Beckman, Corning and Coleman B type electrodes and all others with pin plug connectors.

23 WISSENSCHAFTLICH-TECHNISCHE WERKSTATTEN GmbH

Type Cat. No.	Range pH	±mV	Accuracy	Power	Temp. Comp.	Recorder Output	Special Features	Applications
pH 39	0-8 and 6-14	±500 mV	±0.05pH or ±2mV	Fully mains stabilized, 110, 220 volts, 50,60 C/s	manual 0-100°C	recorder connection provided sensitivity of recorder 40μA, 1700 ohms	Easily transportable, moisture proof and corrosion resistant. Direct reading and very easy to operate.	Suitable for redox measurements and potentiometric and redox titrations and Karl Fischer titrations.
39 i	as above		as above	as above	as above	as above	As above	Suitable for continuous plant pH monitoring with recorder. Particularly useful for pH recording in sewage treatment plants.
390	0-8 and 6-14	±500 ±1500	0.02pH	Fully stabilized 110, 220 V, 50C/s Voltage for Karl Fischer titrations 300 mV	manual 0-100°C	recorder connection provided sensitivity of recorder 40μA, 1700 ohms		Suitable for all potentiometric laboratory titrations including dead-stop titrations.
pH recorder pH 200	0-14	±700	±0.1 pH or ±6mV	mains independent power for one week rechargeable batteries.	not listed		Water and gas proof, shock resistant and corrosion and flood proof.	Specially suitable for sewer measurements, canals and pre-flooder, also industrially for ambulatory and mains independent plant monitoring
Compact pH 54	2-12 or 0-14		±0.1 pH	batteries	not listed	not listed	Fixed point switching for calibration check. Slope screw for electrode connection.	Useful for direct field measurements, for plant monitoring, sewer monitoring, water testing, biological and agricultural field measurements
Redox Tester pH54/RT	as above		as above	as above	as above	as above	As above and special combined electrode to select pH or redox measurements ±500mV or rH value	
Chemometer PLT 10	2-12	±500	±0.05pH				Cond. 0.3 to 10000 μ mhos 1000c/s/±2% $R = 10^6 - 10^{12}$ ohm ± 2-5%	A multi purpose instrument for measurements of pH and mV values conductivity or high resistances.

1 AUTOMATIC TITRATORS

	Beckman Automatic Titrator Model K	Coleman Auto-titrator Model 19	Doran Automatic Titrimeter M. 4910	E.I.L. Automatic Titrimeter Model 24	Pye Automatic Titrator	Radiometer titrator Type TTTI
Mode of Reagent addition	Inter-mittent	Inter-mittent	Contin-uous flow	Continuous flow	Continuous flow	Inter-mittent
Means of delivery	Through a diaphragm valve opened by actuating a solenoid operated arm	Through diaphragm valve solenoid-operated	Through glass valve regulated by solenoid operated needle	All glass tap regulated by solenoid	Through arms of delivery unit regulated by solenoid operated pinching devices on flexible tubes	From an ordinary burette through flexible tubing
Connection to Recorder	Possible	Not listed	Not listed	Not possible	Not listed	Possible
Separation of pH meter and control unit	Possible	Possible	Not possible	Not possible but can function as a null point pH meter at room temp. (no temp. compensation)	Possible	Possible
Burette type	Automatic zero burette			Automatic zero burette		
Special Features	Titration starts when sample beaker is placed in position and stops when it is withdrawn. Circuitry and adjustable electrode position w.r. to burette combine for accurate and point anticipation	After endpoint is reached, instrument continues for pre-selected time then switches off, stirrer locks burette valve & signals con-clusion	Endpoint antici-pation range is 0-2pH or 0-200mV		Titrator consists of control & delivery units connected to one of the stan-dard range of PYE pH meters. Other pH meters may be used if their out-put current is capable of develop-ing 10 V across the input titrator (input resistance 10000 ohms) over 10 pH unit range	Possesses a high zero stability. When endpoint is reached the instrument waits for a preset tine (5-30 secs) before switching off.